MW01488809

Integrated Computational Materials Engineering

A Transformational Discipline for Improved Competitiveness and National Security

Committee on Integrated Computational Materials Engineering

National Materials Advisory Board

Division of Engineering and Physical Sciences

NATIONAL RESEARCH COUNCIL
OF THE NATIONAL ACADEMIES

THE NATIONAL ACADEMIES PRESS
Washington, D.C.
www.nap.edu

THE NATIONAL ACADEMIES PRESS   500 Fifth Street, N.W.   Washington, DC 20001

NOTICE: The project that is the subject of this report was approved by the Governing Board of the National Research Council, whose members are drawn from the councils of the National Academy of Sciences, the National Academy of Engineering, and the Institute of Medicine. The members of the committee responsible for the report were chosen for their special competences and with regard for appropriate balance.

This study was supported by the Department of Defense under Contract No. MDA972-01-D-001 and the National Nuclear Security Administration and the Office of Energy Efficiency and Renewable Energy at the Department of Energy under Contract No. DE-AM01-04P145013. Any opinions, findings, conclusions, or recommendations expressed in this publication are those of the authors and do not necessarily reflect the views of the organizations or agencies that provided support for the project.

Cover: A cast aluminum engine block modeled with an ICME tool. Image courtesy of J.E. Allison, Ford Motor Company. Cover design by Steven Coleman.

International Standard Book Number 13:   978-0-309-11999-3
International Standard Book Number 10:   0-309-11999-5

Available in limited quantities from

National Materials Advisory Board
500 Fifth Street, N.W.
Washington, DC 20001
nmab@nas.edu
http://www.nationalacademies.edu/nmab

Additional copies of this report are available from the National Academies Press, 500 Fifth Street, N.W., Lockbox 285, Washington, DC 20055; (800) 624-6242 or (202) 334-3313 (in the Washington metropolitan area); Internet, http://www.nap.edu.

THE NATIONAL ACADEMIES
Advisers to the Nation on Science, Engineering, and Medicine

The **National Academy of Sciences** is a private, nonprofit, self-perpetuating society of distinguished scholars engaged in scientific and engineering research, dedicated to the furtherance of science and technology and to their use for the general welfare. Upon the authority of the charter granted to it by the Congress in 1863, the Academy has a mandate that requires it to advise the federal government on scientific and technical matters. Dr. Ralph J. Cicerone is president of the National Academy of Sciences.

The **National Academy of Engineering** was established in 1964, under the charter of the National Academy of Sciences, as a parallel organization of outstanding engineers. It is autonomous in its administration and in the selection of its members, sharing with the National Academy of Sciences the responsibility for advising the federal government. The National Academy of Engineering also sponsors engineering programs aimed at meeting national needs, encourages education and research, and recognizes the superior achievements of engineers. Dr. Charles M. Vest is president of the National Academy of Engineering.

The **Institute of Medicine** was established in 1970 by the National Academy of Sciences to secure the services of eminent members of appropriate professions in the examination of policy matters pertaining to the health of the public. The Institute acts under the responsibility given to the National Academy of Sciences by its congressional charter to be an adviser to the federal government and, upon its own initiative, to identify issues of medical care, research, and education. Dr. Harvey V. Fineberg is president of the Institute of Medicine.

The **National Research Council** was organized by the National Academy of Sciences in 1916 to associate the broad community of science and technology with the Academy's purposes of furthering knowledge and advising the federal government. Functioning in accordance with general policies determined by the Academy, the Council has become the principal operating agency of both the National Academy of Sciences and the National Academy of Engineering in providing services to the government, the public, and the scientific and engineering communities. The Council is administered jointly by both Academies and the Institute of Medicine. Dr. Ralph J. Cicerone and Dr. Charles M. Vest are chair and vice chair, respectively, of the National Research Council.

www.national-academies.org

Preface

Integrated computational materials engineering (ICME) is an emerging discipline that aims to integrate computational materials science tools into a holistic system that can accelerate materials development, transform the engineering design optimization process, and unify design and manufacturing. As this report shows, even in its nascent state, developing ICME represents a grand challenge. If ICME is successful, it will provide significant economic benefit and accelerate innovation in the engineering of materials and manufactured products. To that end, the committee that wrote this report was asked by its sponsors at the Departments of Energy and Defense to develop a strategy for the coordinated and accelerated development of this important new technology area. In particular, the committee was charged with the following tasks:

1. The exploration of the benefits and promise of ICME to materials research through a series of case studies of compelling materials research themes that are enabled by recent advances and accomplishments in the field of computational materials.
2. An assessment of the benefits of a comprehensive ICME capability to the national priorities.
3. The establishment of a strategy for the development and maintenance of an ICME infrastructure, including databases and model integration activities. This should include both near-term and long-range goals, likely participants, and responsible agents of change.

4. Making recommendations on how best to meet the identified opportunities.

In executing this charge the committee met four times between November 2006 and October 2007. At its meetings the committee heard from representatives of its sponsors, the Department of Defense and the Office of Energy Efficiency and Renewable Energy and the National Nuclear Security Administration at the Department of Energy. The committee also heard from a broad spectrum of speakers from government, industry, and academia. In particular the committee wants to thank the following people for their contributions to this study: Paul Avery, Brian Baker, Henry Bass, David Benson, Cate Brinson, Frank Brown, Joe Carpenter, Dureseti Chidambarrao, Rex Chisholm, Julie Christodoulou, Leo Christodoulou, Edward Damm, Dennis Dimiduk, Michael Doyle, Marty Fritts, Dave Furrer, Gerry Gibbs, Sharon Glotzer, Martin Green, Robert Hanrahan, Daryl Hess, David Hibbitt, Ursula Kattner, Charles Kuehmann, Dimitri Kusnezov, Kirk Levedahl, Brett Malone, David McDowell, Al Miller, Todd Osman, Ruth Pachter, Robert Pfahl, Krishna Rajan, Nuno Rebelo, Alex Szalay, Louis J. Terminello, Alex Van der Velden, Erich Wimmer, Michael Winter, Gerry Young, and Jonathan Zimmerman. The presentations of these experts helped the committee to build as complete a picture as possible of the current state of this emerging field. The committee's discussions with the presenters and with members of the ICME community and the broader materials engineering community at a town hall meeting in connection with the 2007 annual meeting of The Minerals, Metals & Materials Society (TMS) were key to developing the committee's vision for ICME. The committee is grateful to the leadership of TMS for its support of the town meeting.

My personal thanks also go to the members of the committee for their considerable time commitment and their efforts in writing this report. I am particularly grateful to the vice chair of the committee, John Allison, who provided exceptional leadership and vision and without whom neither the study nor the report would have happened. The committee is also grateful to Michael Moloney of the National Research Council staff, who guided it through the study process.

The committee hopes that this report will inspire the materials community, including the government agencies that support the field, to undertake the tasks it has identified as being important to the successful and timely development of ICME. The committee is convinced that ICME offers significant promise to stimulate new economic development in the United States, as well as to underpin national security and transform the materials profession.

Tresa M. Pollock, *Chair*
Committee on Integrated Computational Materials Engineering

Acknowledgment of Reviewers

This report has been reviewed in draft form by individuals chosen for their diverse perspectives and technical expertise, in accordance with procedures approved by the National Research Council's Report Review Committee. The purpose of this independent review is to provide candid and critical comments that will assist the institution in making its published report as sound as possible and to ensure that the report meets institutional standards for objectivity, evidence, and responsiveness to the study charge. The review comments and draft manuscript remain confidential to protect the integrity of the deliberative process. We wish to thank the following individuals for their review of this report:

Paul Avery, University of Florida,
L. Catherine Brinson, Northwestern University,
Rex Chisholm, Northwestern University,
Anthony G. Evans, University of California, Santa Barbara,
Sharon C. Glotzer, University of Michigan,
George (Rusty) T. Gray III, Los Alamos National Laboratory,
Craig S. Hartley, El Arroyo Enterprises LLC,
David Hibbitt, Abaqus, Inc. (retired),
Paul Mason, Thermo-Calc Software, Inc.,
Roger C. Reed, The University of Birmingham,
David J. Srolovitz, Yeshiva University,
Patrice E.A. Turchi, Lawrence Livermore National Laboratory,

Mark Verbrugge, General Motors, and
James C. Williams, Ohio State University.

Although the reviewers listed above have provided many constructive comments and suggestions, they were not asked to endorse the conclusions or recommendations, nor did they see the final draft of the report before its release. The
review of this report was overseen by Stephen Davis of Northwestern University.
Appointed by the National Research Council, he was responsible for making certain
that an independent examination of this report was carried out in accordance with
institutional procedures and that all review comments were carefully considered.
Responsibility for the final content of this report rests entirely with the authoring
committee and the institution.

Contents

Summary

Materials innovations have been at the core of the vast majority of major disruptive technologies since the start of the industrial revolution. Modern transportation, electronics, space exploration, the information age, and medical prosthetics were all enabled by today's metallic, polymeric, ceramic, semiconductor, and multifunctional materials. For decades, the development of advanced materials and their incorporation into the design of new products enabled the United States to maintain a significant competitive advantage in the global economy. Modern computational engineering tools generally have radically reduced the time required to optimize new products. However, analogous computational tools do not exist for materials engineering. As a result, the product design and development cycle now outpaces the materials development cycle, leading to a considerable mismatch. The insertion of new materials technologies has become much more difficult and less frequent, with materials themselves increasingly becoming a constraint on the design process. The materials development and optimization cycle can no longer operate at the rapid pace required, and this potentially threatens U.S. competitiveness in powerhouse industries such as electronics, automotive, and aerospace, in which the synergy among product design, materials, and manufacturing has given our nation a competitive advantage. Moreover, this deficiency leads to suboptimal materials and engineering solutions to national security needs. Until materials engineering, component design, and manufacturing engineering are integrated, designers will not attempt to optimize a product's properties through processing, and one route to improving the competitiveness of U.S. manufacturers will be closed off.

A new and promising engineering approach known as integrated computational materials engineering (ICME) has recently emerged. Its goal is to enable the optimization of the materials, manufacturing processes, and component design long before components are fabricated, by integrating the computational processes involved into a holistic system. Not only a concept, ICME can also be considered a discipline in that it is taking the formative steps of establishing tools, an infrastructure, methodologies, technologies, and even a community to accomplish this goal. For the purposes of this report, ICME can be defined as the integration of materials information, captured in computational tools, with engineering product performance analysis and manufacturing-process simulation. The emphasis in ICME is on the "I" for integrated and "E" for engineering. Computational materials modeling is a means to this end. The grand challenge for the field of materials science and engineering is to build an ICME capability for all classes and applications of materials.

This report describes how ICME is demonstrating its potential to provide a significant return on investment in a few key materials applications. It also describes how the national security of the United States will be enhanced by accelerating innovation in computational materials engineering. The widespread application of ICME promises to transform both the field itself and how the field interacts with the larger engineering process. The impact of ICME on materials engineering will be similar to the impact of bioinformatics on molecular biology. The application of the ICME paradigm to large numbers of materials and manufacturing processes represents a grand challenge for materials engineering, but because it is a very new discipline, success is by no means guaranteed.

ICME is a technologically sound concept whose economic benefits the committee reports on by means of case studies. However, from an industry-wide perspective, ICME is not mature and is still contributing only peripherally to the bottom line. Many in the materials community and in industry are not even aware of its existence or its promise. Nevertheless, ICME shows significant potential for reducing component design and process development costs and cycle times, lowering manufacturing costs, improving the prognosis for material and component life, and, ultimately, allowing for agile response to changing market demands. But for ICME to succeed, three things must happen:

- ICME must be embraced as a discipline in the materials science and engineering community, leading to changes in education, research, and information sharing.
- Industrial reluctance to accept ICME must be overcome. Acceptance is hindered by the slow conversion of science-based computational tools to engineering tools, a lack of awareness and investment, and a shortage of trained computational materials engineers.

- The government must give coordinated support for the initial development of ICME tools, infrastructure, and education. These are currently inadequate yet are critical for ICME's future.

For ICME to become mature, a considerable investment from government and industry will be required. The evidence gathered in this study suggests that there will be a substantial return on this investment.

A number of lessons learned from the early applications of ICME are documented in this report:

- ICME is an emerging discipline, still in its infancy.
- There is clearly a positive return on investment in ICME.
- Achieving the full potential of ICME requires sustained investment.
- ICME requires a cultural shift.
- Successful model integration involves distilling information at each scale.
- Experiments are key to the success of ICME.
- Databases are the key to capturing, curating, and archiving the critical information required for development of ICME.
- ICME activities are enabled by open-access data and integration-friendly software.
- In applying ICME a less-than-perfect solution may still have high impact.
- Development of ICME requires cross-functional teams focused on common goals or "foundational engineering problems."

Realizing the promise of ICME will require technological and cultural challenges to be overcome. The properties of materials are controlled by a multitude of structural features, defects, and often competing mechanisms that operate over a wide range of length scales and timescales. There is no single overarching approach for modeling the entire spectrum of relevant material phenomena. While ICME is now feasible because of the progress made in science-based models and simulation tools, technical and computational gaps exist and will persist for some time. Rapid, targeted experiments and empirical models will be required to fill the gaps where theory is not sufficiently predictive or quantitative. To efficiently create and utilize accurate and quantitative ICME tools, engineers must have easy access to relevant, high-quality data from experimentation and computation. Currently there is much waste in the R&D and engineering system, where materials data are frequently generated and then routinely lost, necessitating redevelopment. Publicly available repositories of high-quality, precompetitive data will accelerate the development of an ICME capability.

An ICME infrastructure will be the enabling framework over which ICME can take place. Some of the elements of that infrastructure (sometimes called a cyberinfrastructure) are libraries of materials models and software tools, integrating software tools, computational hardware, and human expertise. Materials databases (and associated classification schemes or taxonomies) that are openly accessible to the materials development community are essential for ICME. Such databases will allow archiving and mining the large qualified and standardized data sets that are required for development of ICME tools.

For ICME to succeed as a discipline, it must be embraced broadly by the international materials science and engineering community, leading to changes in education, research, and information sharing. The engineering culture must increase its confidence in and reliance on computational materials models as substitutes for large, experimental databases and iterative physical prototypes. Many existing theories, models, and tools, while not perfect, are sufficiently well developed that they can be used for ICME. Sensitivity studies, understanding of real-world uncertainty, and experimental validation are key to gaining acceptance for and value from ICME tools that are less than 100 percent accurate.

This report presents the committee's analysis of the current status of ICME and describes the barriers to its development as well as ways to overcome them. It proposes a strategy to promote the development of ICME by defining actions for the stakeholders involved. It is in this context that the committee offers the following recommendations.

RECOMMENDATIONS

Recommendation 1: As part of its critical mission to advance the nation's economic and energy security, the Department of Energy (DOE) should pursue the following actions:

- **The Office of Energy Efficiency and Renewable Energy (EERE), as an early champion of ICME, should continue to take the lead in the automotive sector and to extend the ICME approach to other compelling applications in energy generation and storage technologies.**
- **The National Nuclear Security Administration (NNSA) should build on its success in creating robust computational materials science tools for predicting the long-term behavior of nuclear weapons systems by integrating them into an ICME system and then extending that system when the chance arises to other suitable materials. In the process, the NNSA should critically assess integration issues and establish best practices for the dissemination of ICME tools to the defense and commercial sectors for further application and validation.**
- **The Office of Science's Basic Energy Sciences (BES) should support a critical link within ICME by utilizing its unique facilities to advance rapid materials characterization and to connect new rapid characterization**

techniques with its strong university and national laboratory programs in computational materials science.

- The Office of the Secretary of Energy should establish an intra-agency ICME coordination group to champion development of ICME across DOE in the research programs supported by BES, EERE, and NNSA as well as in the Office of Nuclear Energy, the Office of Fossil Energy, the Office of Fusion Energy Sciences, and the Office of Advanced Scientific Computing Research. One task for the coordination group should be to establish incentives and requirements for materials researchers to incorporate materials information into open-access infrastructures, together with processes to ensure that the information and models can be used effectively.

Recommendation 2: In view of the benefits of ICME to national security, the Department of Defense (DOD) should expand its leadership role as an early champion of ICME and establish a long-range coordinated ICME program that will accomplish the following:

- Identify and pursue at least one key foundational engineering problem in each service to accelerate the development and application of ICME to critical defense platforms and
- Develop an ICME infrastructure of precompetitive material process–structure–property tools and databases for defense-critical systems.

In addition, DOD should establish an intra-agency ICME coordination group to champion development of ICME within the military and the defense industry.

Recommendation 3: The National Science Foundation—through its Office of Cyberinfrastructure, its Directorate of Engineering, and its Division of Materials Research—should

- Fund cross-disciplinary research and engineering partnerships to develop the taxonomy, knowledge base, and cyberinfrastructure required for ICME.
- Establish incentives and requirements for materials researchers to place their materials information in open-access infrastructures, together with procedures to ensure that the information and models can be used effectively.
- Develop engineering talent for ICME by supporting innovative curricula and student internship programs.

Recommendation 4: To promote U.S. innovation and industrial competitiveness, the National Institute of Standards and Technology (NIST) should develop and curate precompetitive materials informatics databases and develop automated tools for updating, integrating, and accessing ICME resources.

Recommendation 5: Federal agencies should direct Small Business Innovation Research (SBIR) and Small Business Technology Transfer (STTR) funding to support the establishment of ICME-based small businesses.

Recommendation 6: In pursuit of the promise of ICME to increase U.S. competitiveness and support national security, the Office of Science and Technology Policy should establish an interagency working group under the Networking and Information Technology Research and Development initiative to set forth a strategy for ICME interagency coordination, including promoting access to data and tools from federally funded research.

Recommendation 7: U.S. industry should identify high-priority foundational engineering problems that could be addressed by ICME, establish consortia, and secure resources for implementation of ICME into the integrated product development process.

Recommendation 8: The University Materials Council (UMC), with support from materials professional societies and the National Science Foundation, should develop a model for incorporating ICME modules into a broad spectrum of materials science and engineering courses. The effectiveness of these additions to the undergraduate curriculum should be assessed using Accreditation Board for Engineering and Technology, Inc., criteria.

Recommendation 9: Professional materials societies should

- Foster the development of ICME standards (including a taxonomy) and collaborative networks.
- Support ICME-focused programming and publications.
- Provide continuing education in ICME.

In the long term—that is, 10-20 years—as a result of coordination and targeted investment by stakeholders in the critical elements of ICME, the report looks to the following transformational vision to be realized: ICME will become a critical element for maintaining the competitiveness of the U.S. manufacturing base. To enable the rapid design and optimization of new materials, manufacturing processes, and products, ICME practitioners—a broad spectrum of scientists, engineers, and manufacturers—will have open access to a curated ICME cyberinfrastructure, including libraries of databases, tools, and models. All researchers and workers in materials and manufacturing will benefit from ICME. Even the materials scientist in academia performing traditional science-based inquiry will benefit from the assembled and networked data and tools. Discoveries will be easier and their translation to innovative engineering products more straightforward. ICME will have reduced the materials development cycle from today's 10-20 years to 2 or 3 years. And graduating materials science and engineering students will be employed and operate in a multidisciplinary and computationally rich engineering environment.

1

A Vision for Integrated Computational Materials Engineering

The economic vitality and security of the United States in the twenty-first century will depend on the rapid design and manufacture of complex engineering systems, as well as the rapid validation of their safety and effectiveness. Key challenges for the nation include developing new technologies for energy security; ensuring the superiority of U.S. defense capabilities, including the safe operability of the U.S. strategic weapons stockpile; and maintaining the U.S. competitive edge in industrial sectors such as aerospace, automotive, biomedical, and communications and information technology.

In the swiftly changing and increasingly competitive global marketplace, innovative design solutions and short product development cycles that rely on integrated product development teams (IPDTs) armed with computationally based design, engineering analysis, and manufacturing tools are what give the nation its competitive advantage. A critical missing link in the integrated product development process is a set of predictive computational *materials* engineering tools. The development of computational tools for materials engineering has lagged behind the development of such tools in other engineering fields because of the complexity and sheer variety of the materials and physical phenomena that must be captured. In spite of these scientific challenges, the computational tools for materials engineering are now reaching the level of maturity where they will have a substantial impact if they can be integrated into the product development process.[1,2] This integration is the basis for the emerging discipline of integrated computational materials engineering (ICME). ICME promises to eliminate the growing mismatch between the materials development cycle and the product development cycle by integrating materials computational tools and information with the sophisticated computational and analytical tools already in use in engineering fields other than materials. ICME will be transformative for the materials discipline, promising to shorten the materials development cycle from its current 10-20 years to 2 or 3 years in the best scenarios. ICME will permit materials to be "design solutions" rather than selections from a static menu. As an emerging discipline, ICME can be usefully defined as follows:

> Integrated computational materials engineering (ICME) is the integration of materials information, captured in computational tools, with engineering product performance analysis and manufacturing-process simulation.

By materials information the committee means curated data sets, structure–property models, processing–structure relationships, physical properties, and thermodynamic, kinetic, and structural information. Figure 1-1 shows an idealized ICME system that brings together the many kinds of materials information needed for product development—for example, system requirements, manufacturing process–material microstructure models, microstructure–property models, materials databases, cost analyses, and models for how material variability can cause uncertainty in performance. This information can be supplied to geographically dispersed design teams on demand and then linked to the computational tools of other engineering disciplines involved in product design and manufacturing, such as finite-element analysis of product performance or manufacturing process simulation.

ICME represents a breakthrough opportunity for the rapid development, implementation, and validation of cost-effective, advanced engineering systems. It offers a solution to the challenge faced by U.S. industry of developing safe and durable engineered products and inserting them quickly at the lowest possible cost. ICME also promises to be an essential element of the solution to engineering problems that are technically complex, extraordinarily expensive, and—in some cases—difficult or impossible to test and validate at the systems level. ICME is an emerging discipline in the sense that it is still taking the formative steps of devel-

---

[1] National Science Foundation (NSF), *Simulation-Based Engineering Science: Revolutionizing Engineering Science Through Simulation.* Report of NSF Blue Ribbon Advisory Panel, May 2006. Available at http://www.nsf.gov/pubs/reports/sbes_final_report.pdf. Accessed March 2008.

[2] NSF, *From Cyberinfrastructure to Cyberdiscovery in Materials Science: Enhancing Outcomes in Materials Research, Education and Outreach 2006.* Available at http://www.mcc.uiuc.edu/nsf/ciw_2006/. Accessed March 2008.

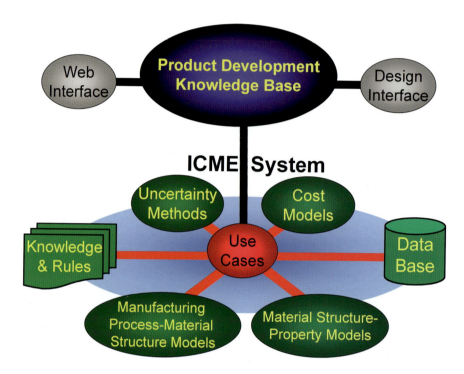

FIGURE 1-1 Schematic structure of an ICME system that unifies materials information into a holistic system that is linked by means of a software integration tool to a designer knowledge base containing tools and models from other engineering disciplines.

oping tools, setting up an infrastructure, methodologies, and technologies, and gathering around it a community.

BENEFITS TO THE NATION

Fundamental changes have occurred in the U.S. and global industrial enterprise over the last 25 years. Successful industries have increasingly focused on lean manufacturing, the elimination of inefficiencies from industrial processes, the rapid adoption of innovative technology, and the globalization of suppliers and customers. There has been a corresponding transformation in engineering, so that today IPDTs simultaneously design high-value systems and establish the processes for fabricating these systems. These multidisciplinary teams have dramatically shortened the product development cycle by using suites of computational design tools that unify formerly disparate technical areas such as heat transfer, aerodynamics, fluid flow, mechanics, electromagnetics, and optics. The result has been that design engineers can focus on the higher-value activity of making decisions based on the output of the integrated design tools rather than collecting and validating data using time-consuming, expensive experimental programs. Additional value is gained through a broader assessment of the parameters important to the product design over a broad spectrum of engineering disciplines and rapid optimization to the best solution; in this sense IPD permits a high-fidelity assessment of what is called the "design space."[3,4] The development and implementation of this integrated capability provides a competitive edge. Not surprisingly, therefore, the market for integration and optimization software is increasing.[5] It is against this backdrop that the promise of ICME is emerging.

Materials are a strategic aspect of engineered products in many different industries, including aerospace, automotive, electronics, and energy generation. Over the years, the development of advanced materials and their incorporation in new products has enabled the United States to maintain a significant competitive advantage in the global economy. Therefore it is a matter of great concern that the materials discipline has not kept pace with the product design and development cycle and that insertion of new materials has become more infrequent.[6,7,8] While the materials engineer is a member of the IPDT, materials selection and materials design now happen outside the computationally driven design optimization loop. As a result, materials are increasingly becoming a design constraint rather than a design enabler. This shortcoming reduces the potential design space, is a drag on innovation, increases manufacturing risk, and gives customers suboptimal end products.[9]

---

[3]Michael Winter, P&W, "Infrastructure, processes, implementation and utilization of computational tools in the design process," Presentation to the committee on March 13, 2007. Available at http://www7.nationalacademies.org/nmab/CICME_Mtg_Presentations.html. Accessed February 2008.

[4]K.G. Bowcutt, "A perspective on the future of aerospace vehicle design," American Institute of Aeronautics and Astronautics Paper 2003-6957, December 2003.

[5]Alex Van der Velden, Engineous, "Use of process integration and design optimization tools for product design incorporating materials as a design variable," Presentation to the committee on March 14, 2007. Available at http://www7.nationalacademies.org/nmab/CICME_Mtg_Presentations.html. Accessed February 2008.

[6]Leo Christodolou, DARPA, "Accelerated insertion of materials," Presentation to the committee on November 20, 2006. Available at http://www7.nationalacademies.org/nmab/CICME_Mtg_Presentations.html. Accessed February 2008.

[7]National Research Council (NRC), *Accelerating Technology Transition: Bridging the Valley of Death for Materials and Processes in Defense Systems*, Washington, D.C.: The National Academies Press (2004).

[8]NRC, *Retooling Manufacturing: Bridging Design, Materials, and Production*, Washington, D.C.: The National Academies Press (2004), p. 53.

[9]The IPDT process is discussed more in Chapter 2. Chapter 4 discusses barriers to ICME implementation, including inertia in the engineering community and in industry.

Conclusion 1: The materials development and optimization cycle cannot operate at the rapid pace required by integrated product development teams, and this potentially threatens U.S. competitiveness in powerhouse industries such as electronics, automotive, and aerospace, in which the synergy among product design, materials, and manufacturing is a competitive advantage.

ICME promises to reinsert materials into the design and manufacturing process optimization loop and give a return on investment (ROI) that will be attributable to a number of factors: design innovation, quicker identification of materials solutions to design problems, faster and less costly new product development, better control of the manufacturing process, and improved capabilities for predicting engineering system performance or life cycle. ICME will allow new products to emerge faster and to achieve a market advantage based on improved performance from incorporating materials and processes optimized for particular applications and on more precise modeling of a material's response to an application environment. ICME can enable the virtual engineering assessment of new materials that might be considered risky to assess with physical prototypes or in systems where the validation of materials performance by system-level testing is expensive, time consuming, or not possible. While its implementation will be a substantial undertaking for both the materials community and the broader engineering community, ICME promises to provide significant economic benefit and will enhance the national security and competitiveness of the United States through accelerating innovation in the engineering of materials and manufactured products.

As described later in this report, ICME case studies have shown early benefit. Examples include the development of nickel-based superalloys for new aeroengine turbine disks[10,11] and the stewardship of the nation's strategic nuclear stockpile, which requires regular updates to 85-year projections on performance and reliability.[12,13] While some of these first demonstrations integrated empirical models

---

[10]NRC, *Accelerating Technology Transition: Bridging the Valley of Death for Materials and Processes in Defense Systems*, Washington, D.C.: The National Academies Press (2004).

[11]NRC, *Retooling Manufacturing: Bridging Design, Materials, and Production*, Washington, D.C.: The National Academies Press (2004), p. 53.

[12]Since the United States continues to observe a moratorium on nuclear testing, the National Nuclear Security Administration (NNSA) has adopted a science-based Stockpile Stewardship Program (SSP) that emphasizes the development and application of greatly improved technical capabilities to assess the safety, security, and reliability of existing nuclear warheads without the use of nuclear testing. One track of this program has been to use an ICME-like approach to investigate the aging of the plutonium. The development and implementation of computational tools were combined with specifically designed validation experiments to satisfy national security requirements within the constraints of existing regulatory and policy limitations.

[13]Louis J. Terminello, Lawrence Livermore National Laboratory, "Synergistic computational/experi-

of materials behavior, the capability will be improved as more predictive science-based numerical models are developed and applied. Early ICME capabilities include the following:

- The efficient exploration of new materials, or variants of existing materials, that satisfy a design constraint;
- The active linking of materials models to explore design trade-offs and permit the optimal exploitation of new material capabilities;
- The optimization at the component level of an improved manufacturing process, decreasing cost and product development time and reducing scale-up risk;
- Reductions in the time and cost of product development; and
- Efficient and accurate forecasting of a material's behavior in service, including performance in environments where validation cannot be accomplished experimentally or requires unrealistic experimental time frames.

The case studies described in Chapter 2 demonstrate that application of ICME, even if in a limited capacity, can result in a significant ROI. Data on such returns reported to the committee varied from one case to another and depended on the class of materials, the expertise required to utilize the ICME tools, and the situation in which the tools were applied. Some of the case studies did not result in a full realization of potential benefits owing to a multitude of factors, including lack of investment and cultural issues. All that notwithstanding, the committee observed that ROIs ranged from 3:1 to 9:1. Not surprisingly, this kind of potential payback from applying the ICME approach is of growing interest in a number of industrial sectors, particularly those in which materials innovations would bring a competitive advantage. However, since a multiyear investment is typically required to build an infrastructure for it, ICME faces a significant challenge in the environment of the 1-year budget cycle that is typical for industry. An important element of ICME successes so far has been the selection of an appropriate foundational engineering problem—that is, a manufacturing process, a material system, and an application or set of applications that steer the development of the computational tools and the infrastructure. Examples of foundational problems that would, if pursued, further accelerate the development of ICME are discussed in Chapter 2.

Conclusion 2: ICME is a technologically sound concept that has demonstrated a positive return on investment and promises to improve

---

mental efforts supporting stockpile stewardship," Presentation to the committee on March 13, 2007. Available at http://www7.nationalacademies.org/nmab/CICME_Mtg_Presentations.html. Accessed February 2008.

the efficient, timely, and robust development and production of new materials and products.

The benefits of ICME are substantial for the nation and its industrial base. However, these benefits will be realized by those industries and nations that develop a proficiency in its application, not by those that restrict access to its building blocks. Like most disciplines, the knowledge infrastructure in materials science and engineering is global. In many cases, significant specialized knowledge resides outside the United States. Moreover, given the complexity and breadth of the activity that is required to realize the vision of ICME, there is a need for international cooperation to accelerate progress and minimize the global investment. Although it did not exhaustively investigate the matter, the committee found evidence of ICME activities in other nations.[14,15] This is a classic cooperation/competition situation, in which resources must be leveraged to develop a basic capability while ensuring a competitive and security advantage by becoming proficient in the application of these tools and by being early adopters that arrive at innovative solutions. While national security interests will dictate that some classes of materials and processing information must be restricted, the committee believes that it will be in the best interest of the nation for ICME and its basic building blocks to be as accessible as possible and available to the widest possible audience. Such building blocks would include all of the ICME cyberinfrastructure, including collaborative Web sites and repositories of data and models. While export control laws have an important purpose, their unnecessary expansion to include all elements of ICME could substantially increase the time and cost of developing a widespread ICME capability and could limit the ability of U.S.-based corporations with a global reach to obtain maximum value from ICME.

CRITICAL ELEMENTS FOR ICME DEVELOPMENT

As discussed throughout this report, while ICME promises to strengthen materials and manufacturing involvement in the integrated product development process, the broad implementation of the ICME paradigm requires significant scientific, computational, and cultural elements to be in place. The widespread development and use of ICME will require a high level of technical maturity for the computational tools, education of science and engineering practitioners in ICME

---

[14]P. Li, D.M. Maijer, T.C. Lindley, and P.D. Lee, "A through process model of the impact of in-service loading, residual stress, and microstructure on the final fatigue life of an A356 automotive wheel," *Materials Science and Engineering A.* 460-461 (July 2007): 20-30.

[15]J. Hirsch, *Virtual Fabrication of Aluminum Products: Microstructural Modeling in Industrial Aluminum Fabrication Processes,* Weinheim, Germany: Wiley-VCH (2006).

capabilities, and confidence on the part of customers and regulatory entities in the outcomes of ICME implementations.

Conclusion 3: While some aspects of ICME have been successfully implemented, ICME as a discipline within materials science and engineering does not yet truly exist.

The committee identified several critical elements as being necessary for widespread ICME development and implementation. The technical challenges clearly require advances in models, infrastructure, and data. Of more importance are the cultural and organizational issues that create significant barriers for the adoption of ICME. The current state and the desired future state of the technical, cultural, and organizational aspects of ICME are described in detail in Chapters 3 and 4, respectively. The grand challenge for materials science and engineering is to build an ICME capability for all classes and applications of engineering materials.

Cultural and Organizational Elements

Perhaps the greatest challenge facing the development of ICME is a cultural one—that is, becoming an established practice in the science and engineering profession. To realize the benefit of ICME to the nation, there must be a change in culture in the academic, industrial, and government materials engineering and R&D organizations. In the committee's judgment, these changes will be fundamental in character and will need to be embodied throughout the materials enterprise.

Acceptance by the Materials Science and Engineering Community

Materials science first emerged as an academic discipline with a strong emphasis on metallurgy in the late 1950s. By the 1970s, as the palette of engineering materials began to expand rapidly, the discipline was defined to be "concerned with the generation and application of knowledge relating the composition, structure, and processing of materials to their properties and uses."[16,17,18] In the 1990s the field, more typically referred to as materials science and engineering, was reaffirmed to cover four topics: (1) properties, (2) performance, (3) structure and composition,

---

[16]R.W. Cahn, *The Coming of Materials Science*, Pergamon (2001).

[17]NRC, *Materials and Man's Needs: Materials Science and Engineering—Volume I, The History, Scope, and Nature of Materials Science and Engineering*, Washington, D.C: National Academy Press (1974). Available at http://books.nap.edu/catalog.php?record_id=10436. Accessed February 2008.

[18]NRC, *Materials Science and Engineering for the 1990s: Maintaining Competitiveness in the Age of Materials*, Washington, D.C: National Academy Press (1989). Available at http://www.nap.edu/openbook.php?isbn=0309039282. Accessed February 2008.

and (4) synthesis and processing. Materials science in its infancy focused on structure and the development of techniques for quantifying structure and defects.[19] It is only recently that modeling and simulation have started to become an accepted and useful part of the materials field. To date, however, computational efforts have been generally focused on the science of materials—that is, on seeking fundamental understanding of materials behavior or discovering new materials—and have not, in general, been transferred to the materials-development and engineering processes.[20,21,22] The true impact of ICME will be realized only when materials science and engineering is fully integrated into the integrated product development process. When that happens, the materials engineer, who will be able to design and create new materials in an integrated way, will play a role like that of the other engineers in an IPDT. In addition, ICME also allows the designer to provide input on the development of new materials, enabling the efficient prioritization of materials development and characterization activities.

In this scenario, the materials engineer exercises computational resources that permit materials selection; prediction of property changes for new component geometries or processing paths for a fixed materials system; prediction of a spectrum of properties for evolutionary versions of a material; and guiding the design of completely new materials. Few materials engineers, however, receive sufficient background in computation or the basics of modeling and simulation in the course of their education to be effective ICME practitioners without additional training. Thus as discussed in more detail in Chapter 4, the undergraduate and graduate curricula will have to undergo major change to prepare materials scientists and engineers for the widespread adoption of ICME, and continuing education will be required to expand the skills of practicing engineers. This change, in turn, will require a change in how the materials community operates.

There is generally a separation between science and engineering in the materials research community, with most academic researchers being focused to a greater degree on science. This science focus is partly due to the nature of the funding sources and partly to a historical bias in academic materials departments. ICME will require and promote a better connection between the science of materials and the engineering of materials. It requires science to provide the fundamental understanding needed to develop better models for ICME. In turn, ICME provides a "market" for that science, allowing it to have a greater impact on materials engineering and materials development processes. This symbiotic relationship will

---

[19]C.S. Smith, *A Search for Structure*, Cambridge, Mass.: MIT Press (1981).

[20]G.B. Olson, *Science* 288 (5468): 993 (2000).

[21]J. Greeley and M. Mavrikakis, *Nature Materials* 3, 810 (2004).

[22]Department of Energy (DOE), "Opportunities for Discovery: Theory and Computation in the Basic Energy Sciences," Office of Science (January 2005). Available at http://www.sc.doe.gov/bes/reports/files/OD_rpt.pdf. Accessed February 2008.

require a shift in how scientists convey the product of their work. The development of ICME as a discipline within materials science and engineering will require data and information sharing on a scale unknown today. No single organization has all the tools or data for the robust application of ICME. Nor do all the tools reside in one country. New paradigms are needed for sharing information and tools and for building an international ICME community. ICME can provide the platform for materials science to connect better to materials engineering.

> **Conclusion 4: For ICME to succeed, it must be embraced as a discipline by the materials science and engineering community, leading to requisite changes in education, research, and information sharing. ICME will both require and promote a better connection between the science of materials and the engineering of materials. ICME will transform the field of materials science and engineering by integrating more holistically the engineering and scientific endeavors.**

Acceptance by Industry

Relative to engineering tools such as computational fluid dynamics and finite element methods, ICME is at the very beginning of its integration into the design process. The continued paucity of computational materials engineering tools that can contribute to the industrial design process at the same level as those other more mature computational engineering tools only increases skepticism about the feasibility of integrating materials tools into the IPDT process. This lack of maturity also leads to concerns about whether materials tools can be validated to the level of fidelity required by regulatory agencies. These concerns constitute major cultural barriers to industry's widespread acceptance of ICME that are as difficult to overcome and as important to address as the technical challenges. These cultural barriers seem especially acute in traditional manufacturing settings, where the experience base of the engineers has an overwhelming influence on critical technical decisions. That base has limited experience with or awareness of ICME. Ensuring the future success of ICME will require a long-term investment, first to develop computational tools that can be integrated with design and manufacturing and then to train materials experts in their use and potential. For the foreseeable future, ICME tool sets will be largely developed by teams of experts with detailed and fundamental materials knowledge for particular materials systems, an understanding of the requirements of the particular engineering component or system of interest, the limits of practical computing tools, and the constraints of engineering time lines. This expertise will be required to build tool sets that allow robust predictive capability while ensuring that computational simulations are sufficiently rapid that the results can be used to impact engineering decisions.

Conclusion 5: Industrial acceptance of ICME is hindered by the slow conversion of science-based materials computational tools to engineering tools and by the scarcity of computational materials engineers trained to use them.

Government Ownership

While there have been several successful government-supported ICME programs—the Accelerated Insertion of Materials (AIM) program at the Defense Advanced Research Projects Agency (DARPA), the Dynamic 3D Digital Structure program at the Office of Naval Research (ONR), and the Advanced Simulation and Computing (ASC) program on materials aging at the DOE's National Nuclear Security Administration (NNSA)—there is no systematic, sustained, coordinated government research program to develop the tools and infrastructures needed for ICME. Coordination is lacking both within and between agencies. This lack of coordination means that ICME tool development is spotty and sporadic. There is duplication of effort, and advances in one arena are not readily available to other researchers or the engineering organizations likely to implement the tools. In the committee's judgment, given the importance of research and development in the materials discipline to the future of defense platforms, energy security, health care, and, ultimately, economic competitiveness, government investment in ICME and coordination of its efforts would have a substantial benefit. Moreover, given the size of the investment required and the worldwide extent of the materials profession, this effort would benefit from global cooperation. A U.S. government cooperative initiative was recommended by previous National Academies panels. A report on accelerated technology transition recommended the establishment of a national, multiagency initiative in computational materials engineering to address three broad areas: methods and tools, databases, and dissemination and infrastructure.[23] Similarly, a report on retooling manufacturing recommended that DOD should create, manage, and maintain open-source, accessible, peer-reviewed tools and databases for materials properties to be used in product and process design simulations.[24] These recommendations do not yet appear to have been acted upon. One possible reason is the absence of a clear framework for accomplishing the goals. With the successful application of ICME to several challenging engineering

---

[23]NRC, *Accelerating Technology Transition: Bridging the Valley of Death for Materials and Processes in Defense Systems*, Washington, D.C.: The National Academies Press (2004).
[24]NRC, *Retooling Manufacturing: Bridging Design, Materials, and Production*, Washington, D.C.: The National Academies Press (2004).

problems and as ICME continues to attract interest, the committee believes that substantial progress is now possible if key stakeholders act.

Conclusion 6: A coordinated government program to support the development of ICME tools, infrastructure, and education is lacking, yet it is critical for the future of ICME.

Technical Challenges

The widespread adoption of ICME approaches will require the development of models and integration tools as well as major efforts in the calibration and validation of models for specific materials systems. Continued evolution and maturation of computational materials science tools will facilitate the introduction of ICME tools. While elements of a comprehensive ICME system exist, significant infrastructural development will be required to realize the benefits of integration and widespread use of ICME in engineering product development. The fundamental technical challenge of ICME is that the materials properties that are essential for design and manufacture involve a multitude of physical phenomena, and that accurately capturing their representation in models requires spanning many orders of magnitude in length scale and time.

Models

From atomistic simulations to finite-element simulations of complex manufacturing operations, the ability to model materials behavior has increased enormously. That said, considerable challenges remain to create processing–structure–property models for materials that can be validated against experiment and then applied across a spectrum of manufacturing conditions. Advances are needed in the basic models themselves, particularly in quantifying the connections among material structure, defects, and material properties in descriptions that are relevant to engineering. Truly predictive capabilities will require materials codes to be compatible with modern computing platforms having multiprocessor and parallel processing capabilities. Other developments that are critical for ICME include the ability to link models together, which for the most part has not been addressed, and then understanding how uncertainty in the models propagates throughout the ICME process. Without a quantified uncertainty, the acceptance of ICME will be limited in many technologically advanced industries.

Conclusion 7: Although there has been significant progress in the development of physically based models and simulation tools, for many key areas they are inadequate to support the widespread use of

ICME. However, in the near term, ICME can be advanced by use of empirical models that fill the theoretical gaps. Thus experimental efforts to calibrate both empirical and theoretical models and to validate the ICME capability are paramount.

Integration Tools

Developing an integration infrastructure that permits multidisciplinary analysis, collaborative model development, and design optimization with materials as a key optimization parameter will be critical for the future growth of ICME. This infrastructure, perhaps more appropriately referred to as the cyberinfrastructure,[25] is composed of many enabling pieces, as discussed in more detail in Chapter 3. These include libraries of computational materials science models and tools, databases, computing capability, complementary experimental tools, and integration and collaboration software. These objects can be local or geographically dispersed and may comprise a mixture of precompetitive and proprietary materials information and models.[26] Often integration will be accomplished via the Internet on collaborative Web sites and in information repositories that are important elements of the cyberinfrastructure. A well-constructed infrastructure will allow single- or multiple-application users and multidisciplinary users to perform collaborative design. Some portions of the infrastructure will be industry specific, while others will span the entire ICME community. Developing a comprehensive ICME infrastructure for all industries critical to U.S. competitiveness will be a major undertaking. Success will require contributions from a spectrum of stakeholders, including industry, government, national laboratories, professional societies, and educational institutions. The challenge arises from the different missions and business models of the stakeholders, who face cultural barriers specific to their business. ICME as a discipline must define mechanisms and resources for collaboration to accomplish the ultimate goal of establishing an integrated set of materials tools. Examples from other disciplines, including biology, demonstrate that the involvement of a broad spectrum of potential users in the early stages of infrastructure development is essential for building a fully functional infrastructure and a strong community of practitioners.

Developing models and databases that can be interfaced into an integration scheme is a major technical requirement for ICME. Researchers in materials

[25]For more information, see NSF, *Revitalizing Science and Engineering Through Cyberinfrastructure: Report of the National Science Foundation Blue Ribbon Advisory Panel on Cyberinfrastructure* (January 2003). Available at http://www.nsf.gov/od/oci/reports/atkins.pdf. Accessed February 2008.

[26]By "precompetitive" the committee means a nonproprietary product in the early stages of development on which competitors might collaborate.

science, materials engineering, physics, and chemistry explore the processing–structure or structure–property relationships of materials and incorporate these findings into sophisticated modeling methods as a natural part of their research; however, they generally focus on a narrow part of the overall materials behavior spectrum. While these sometimes disjointed approaches do not by their nature necessarily contribute to an ICME infrastructure, they represent a vast array of methodologies that can be drawn on by the as-yet-to-be-developed integration framework and software that will be the backbone of ICME.

> **Conclusion 8: An ICME cyberinfrastructure will be the enabling framework for ICME. Some of the elements of that cyberinfrastructure are libraries of materials models, experimental data, software tools, including integration tools, and computational hardware. An essential "noncyber" part of the ICME infrastructure will be human expertise.**

Databases

One of the lessons learned from ICME efforts to date is the profound importance of experimental methods and data to fill gaps in theoretical understanding and validate models. For an ICME strategy to be successful, a strong link between experimental data and modeling is essential. For that data to be accessible to the community, a set of common, open-access databases is needed, in much the same way as the genetics community requires a database of gene sequences. The challenge of providing useful databases of materials information for ICME is that the data can take many forms, depending on the materials system. One of the critical issues in all materials systems, and one that differentiates materials from other disciplines, is how to represent the three-dimensional distribution of a material's microstructure—that is, the three-dimensional distribution of interior features (grain boundaries, phases, defects) in the system. The details of the structure at nano, micro, and higher-order scales have a strong influence on material properties and performance, and a truly predictive material model must account for these features. How to capture and classify three-dimensional microstructural information, as well as the wide range of other data, is an ongoing effort. Having data, from both experiment and modeling, that are as widely available as possible will be of critical importance for the successful application of ICME. Materials development also requires an understanding of how different features in the data may be correlated with other material characteristics. Given the multidisciplinary nature of ICME, an important element of integration efforts will be the development of taxonomies to establish a controlled vocabulary. As discussed in this report, the committee believes that a new field, materials informatics, based on ideas from the biological community, will enable those connections.[27] Advanced materials informatics will enable the ICME expert to find and connect important information when ICME tools are being developed for particular applications. An important element of materials informatics will be establishment of a widely agreed-on taxonomy for describing and classifying materials information.

Conclusion 9: Creation of a widely accepted taxonomy, an informatics technology, and materials databases openly accessible to members of the materials research and development, design, and manufacturing communities is essential for ICME.

Rapid, Targeted Experimentation and Three-Dimensional Characterization

The availability in ICME databases of pedigreed data on underlying physical properties and structural information is critical to the development of high-level material property models. While the research and development literature contains a great deal of useful information that could be harvested, newly developed materials or new processing routes inevitably call for experimental data for the calibration and validation of models. There are emerging suites of new characterization tools that permit materials properties to be rapidly screened and evaluated without the need for large volumes of material. Among the new techniques are local laser-based probes for thermal and electrical conductivity, thin-film combinatorial processing for property evaluation, microscale mechanical tests, and the rapid generation of phase diagrams.[28] Also, for probing structural features or defects that are irregular or exist in larger volumes of material, three-dimensional materials tomography tools are also being developed at various length scales.[29,30,31] These techniques and their widespread applications are still in the formative stages, and protocols for acquiring, storing, and sharing the vast amounts of data that might be generated by these new techniques have yet to be developed.

[27]For more information on materials informatics, see http://www.tms.org/pubs/journals/jom/0703/peurrung/peurrung-0703.html. Accessed December 2007.

[28]J.C. Zhao, "Combinatorial approaches as effective tools in the study of phase diagrams and composition–structure–property relationships," *Progress in Materials Science* 51: 557–631 (2006).

[29]J.E Spowart, H.M. Mullens, and B.T. Puchala, *JOM (Journal of The Minerals, Metals & Materials Society)* 55(10): 35 (2003). Available at http://www.aps.anl.gov/Science/Highlights/2001/microtomography.htm. Accessed March 2008.

[30]M.K. Miller, *Atom Probe Tomography: Analysis at the Atomic Level*, Springer (2000).

[31]A.J. Kubis, G.J. Shiflet, D.N. Dunn, and R. Hull, "Focused ion-beam tomography," *Metallurgical Materials Transactions* 35, 1543 (2004).

Conclusion 10: The development of rapid characterization tools alongside new information technology and materials databases will allow speedy calibration of the empirical models required to fill gaps in theoretical understanding.

GOALS AND MILESTONES

Widespread development and application of ICME promises to transform the materials field and how it functions in relation to the engineering process. Molecular biology and medicine are currently undergoing such a transformation in the wake of bioinformatics tools and databases. Development and application of an ICME capability for a large number of materials and manufacturing processes represents a grand challenge for the materials field. However, although ICME could contribute greatly to the security and economic vitality of the United States, the materials community does not yet see it as a discipline. During the course of the study the committee became convinced that now is a critical time for ICME. To continue to provide the strategic advantage that materials engineering has traditionally provided to advanced engineering systems, the materials field must advance its computational capability to match the capabilities in other fields of engineering. This effort will require long-term vision and coordination as well as some short-term actions on the part of government, industry, academic institutions, and materials professional societies. To provide guidance on how the U.S. research and industrial infrastructure can make progress in developing the critical elements—technical, cultural, and organizational—of ICME described above, the committee has identified some short-term goals—milestones for development—that will propel ICME toward maturity in the next 10 years or so. Passing these milestones is the foundation of the strategy the committee identified for the development of ICME and the associated recommendations.

Conclusion 11: To set ICME on the right course of development and to allow it to realize its promise by 2020, the following technical milestones must be passed and programs and activities must be under way:

Tools and Technical Advances

—**Automated tools to access and update existing materials databases.**
—**A core set of science-based processing–structure–property codes that exploit parallel computational processing and are designed for integration and interoperability.**
—**A protocol for translating published data and models into ICME tools.**

—Materials taxonomies and imaging standards for open-access databases.

—An open-access ICME integration and collaboration platform for model development.

—Innovative rapid characterization and three-dimensional imaging techniques and protocols for capturing and sharing these data.

—Uncertainty models for materials properties and performance.

Programs and Activities

—Coordinated ICME research programs at the federal research support agencies.

—Precompetitive industry-led consortia that identify ICME needs and drive development of ICME models and tools.

—An interagency working group to assess the value of ICME in pursuit of national priorities, identify foundational engineering problems, and establish cooperative programs between agencies.

—A program to exploit and demonstrate the potential of ICME by solving at least 10 diverse foundational engineering problems from different industries.

—Small Business Innovation Research (SBIR) and Small Business Technology Transfer (STTR) programs on ICME to support small suppliers of ICME technology.

—A model curriculum and curriculum modules that integrate ICME tools into a broad range of materials science and engineering courses.

—Support from materials professional societies and academic institutions to ensure that ICME is recognized as an emerging discipline.

—Documentation and publication of successes and failures so that others may learn about opportunities and needs and help to build an ICME community.

Identifying and then beginning to solve some foundational engineering problems could be the first steps in developing and demonstrating an ICME framework. A foundational engineering problem consists of an advanced engineering component, a materials system, and a manufacturing process that must be rapidly optimized within a more complex engineering system. Some examples of foundational engineering problems include these:

- High-dielectric materials and processes for improving the performance of microelectronic devices,
- Low-cost organics for robotics sensors,
- Thermal protection materials for hypersonic vehicle surfaces,
- Catalysts for optimizing the performance of hydrogen-fueled systems,
- Reliable and rapid recertification of components in aging structures,
- Materials for ballistic and blast survivability of ship hulls,
- Thermoplastic injection-molded materials for automotive structures,
- Materials and electrochemical processes for advanced batteries,
- Nanoparticles for magnetic storage devices, and
- Composite or advanced metallic materials for aeroengine components.

All these topics and foundational engineering problems could benefit by integrating material structure, property, and process models with the rapidly evolving engineering requirements, resulting in high-performance engineered components. To make real progress in developing ICME tools and demonstrating their capabilities requires a sizeable investment—for example, $10 million to $40 million per program. In the face of the constrained budgets of typical government programs, these foundational engineering problems would have to be further refined and limited to specific material systems, manufacturing processes, and component families. To demonstrate this point, DARPA's Accelerated Insertion of Materials (AIM)[32] program would be typical of the last item in the above list. AIM made important progress on a foundational engineering problem by developing and integrating a suite of process, microstructure-property, and uncertainty models to optimize nickel-based alloy engine disks manufactured by forging. Making meaningful progress in tackling foundational engineering problems such as those listed above will require a similar degree of refinement and focus on specific systems. Selecting a specific set of materials systems, manufacturing processes, and components for a foundational engineering problem would require a detailed knowledge of the priorities and opportunities in each industry and funding agency and would, therefore, be outside the scope of this study.

The committee believes that the milestones and programs listed in Conclusion 11 constitute the requirements for the successful development of ICME over the next dozen years. The recommendations that follow assign responsibilities to various actors in the private and public sectors that will play a role in developing ICME and passing these milestones.

---

[32]For more information on the AIM program, see http://www.darpa.mil/dso/thrusts/matdev/aim/overview.html. Accessed February 2008.

STRATEGY FOR ICME DEVELOPMENT: RECOMMENDATIONS

The development of ICME will require the active participation of a diverse collection of stakeholders—including government, the national laboratories, industry, academia, and the professional societies. By acting on systematic plans for ICME development, these stakeholders could enhance U.S. security and competitiveness. Chapter 4 discusses in much more detail the various roles described in the recommendations below. Figure 1-2 shows these stakeholders and the goals that will have to be reached to achieve the vision of ICME.

Government Role

Because materials are a key element of many advanced engineering systems, there are multiple government stakeholders in ICME. The committee concluded that the development of ICME will require coordination and the sustained effort of a number of government research agencies. Unlike industry, the government is not dominated by the need to meet near-term financial objectives and can have a longer-term perspective on research objectives. The agencies of the federal government that support research—in particular DOD and DOE—play the critical role of identifying and prioritizing topics for investigation. The National Science Foundation (NSF), the National Institute for Standards and Technology (NIST), and DOE's Office of Basic Energy Sciences (BES) play critical roles in developing and disseminating the supporting fundamental science databases, informatics, and cyberinfrastructures. The committee concludes that for the United States to develop a valuable and productive ICME infrastructure in a timely manner, each of these agencies must establish long-range ICME programs and coordination offices to support the development of ICME tools and infrastructures around specific high-priority materials systems and/or defense platforms. Applying ICME to several high-impact applications would motivate a preliminary set of tools and, importantly, further development that could be sustained on a commercial basis. Based on experience to date, each ICME foundational engineering problem might require $10 million to $40 million of total effort over 3 to 10 years, depending on the complexity and the level of completeness desired. Absent major and sustained government coordination and support, ICME development will be slow and uncertain, driven more by grass roots researchers, individual companies, consortia, and professional societies; however, in this scenario, opportunities to insert ICME into the product development process of the major U.S. industries would be missed. At the same time, because ICME is at such an early stage of development, it is important that the coordination within and across agencies working on ICME not be prescriptive and that it allow pursuing alternative approaches to particular problems with the expectation that the strongest approaches for particular applications

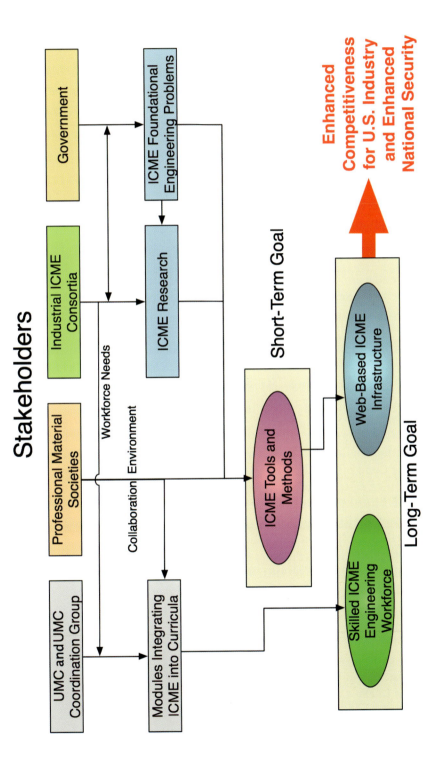

FIGURE 1-2 Overview of the strategy for ICME development that identifies stakeholders and short- and long-term goals. NSF, National Science Foundation; UMC, University Materials Council.

will emerge and can then be developed. Coordination will also ensure that the correct balance in the portfolios of the agencies supporting ICME can be maintained, including the balance between fundamental and applied research. The committee recognizes that because many of the barriers to ICME are not in basic science, the goals of ICME could not be accomplished by simply redirecting support away from basic research to applied research.

Department of Energy

Having concluded that ICME has significant untapped potential to provide a capability for designing new materials for use in energy production and improving the efficiency of its storage and use, the committee offers the following recommendation to DOE:

Recommendation 1: As part of its critical mission to advance the nation's economic and energy security, the Department of Energy (DOE) should pursue the following actions:

- **The Office of Energy Efficiency and Renewable Energy (EERE), as an early champion of ICME, should continue to take the lead in the automotive sector and to extend the ICME approach to other compelling applications in energy generation and storage technologies.**
- **The National Nuclear Security Administration (NNSA) should build on its success in creating robust computational materials science tools for predicting the long-term behavior of nuclear weapons systems by integrating them into an ICME system and then extending that system when the chance arises to other suitable materials. In the process, the NNSA should critically assess integration issues and establish best practices for the dissemination of ICME tools to the defense and commercial sectors for further application and validation.**
- **The Office of Science's Basic Energy Sciences (BES) should support a critical link within ICME by utilizing its unique facilities to advance rapid materials characterization and to connect new rapid characterization techniques with its strong university and national laboratory programs in computational materials science.**
- **The Office of the Secretary of Energy should establish an intra-agency ICME coordination group to champion development of ICME across DOE in the research programs supported by BES, EERE, and NNSA as well as in the Office of Nuclear Energy, the Office of Fossil Energy, the Office of Fusion Energy Sciences, and the Office of Advanced Scientific Computing Research. One task for the coordination group should be to**

establish incentives and requirements for materials researchers to incorporate materials information into open-access infrastructures, together with processes to ensure that the information and models can be used effectively.

EERE became an ICME champion when it funded the first ICME consortium within the U.S. automotive industry.[33,34] The committee believes there is potential for a tremendous return on investment should EERE decide to play a larger role in ICME by championing efforts in other materials areas having an impact on energy production and energy efficiency. DOE's NNSA played a key role in developing the computational materials science tools that have been effectively used for predicting the long-term behavior of a narrow range of materials in nuclear weapons systems. Those efforts could be extended, the committee is convinced, to the comprehensive set of materials utilized in weapons systems and integrated into an ICME framework for nuclear weapons programs. The committee concludes that NNSA laboratories have a unique opportunity to develop such capabilities and to export the ICME framework to the wider materials community. BES could also play an important role in the development of ICME by working on the core fundamental science, computational models, and theory and could leverage its efforts in rapid characterization of materials. It could also play a role in the development of materials informatics and of the ICME cyberinfrastructure.

Department of Defense

DOD is responsible for developing and deploying advanced weapons systems of extreme complexity. Because advanced materials are foundational to the performance of those weapons systems, ICME could offer a critical technological advantage. DOD sponsored some of the first ICME activities. The separate programs of DARPA, the Air Force, and the Navy, described in Chapter 2, were groundbreaking efforts to develop the ICME capability and to build an awareness of ICME and its benefits. DOD has an opportunity to maintain its leadership and to leverage ICME capability for the enhancement of national security by providing focused product development objectives and long-term sustained investments. While this will probably require a slight increase in near-term funding, there is significant potential for

---

[33]Joseph Carpenter, DOE, "DOE's work on extruded long-fiber-reinforced polymer-matrix composites," Presentation to the committee on March 13, 2007. Available at http://www7.nationalacademies.org/nmab/CICME_Mtg_Presentations.html. Accessed February 2008.

[34]The goals of the EERE ICME program are (1) the establishment of a user-friendly, globally accessible ICME for a magnesium cyberinfrastructure and (2) support for multiscale simulations, support for design optimizations under uncertainty, and access to remote databases and repositories of codes.

long-term efficiency improvements. In a time of increasingly constrained funding for DOD materials research, ICME would be one way to improve the efficiency of developing new materials systems, in terms of both cost and timing.

> **Recommendation 2: In view of the benefits of ICME to national security, the Department of Defense should expand its leadership role as an early champion of ICME and establish a long-range coordinated ICME program that will accomplish the following:**
>
> - **Identify and pursue at least one key foundational engineering problem in each service to accelerate the development and application of ICME to critical defense platforms and**
> - **Develop an ICME infrastructure of precompetitive material process–structure–property tools and databases for defense-critical systems.**
>
> **In addition, DOD should establish an intra-agency ICME coordination group to champion development of ICME within the military and the defense industry.**

The committee concluded that given the many overlaps among the materials needs of the military services, coordination is needed to maximize the value and minimize duplication of effort. The tasks for the DOD ICME coordination group could include these:

- Identification of DOD ICME needs, including budget requirements.
- Establishment of a long-range (15-20 year) strategy and a DOD-specific roadmap for funding, developing, and implementing ICME.
- Identification and pursuit of some initial priority foundational engineering problems associated with defense platforms that could substantially benefit from new ICME capabilities.
- Funding for the development of an ICME taxonomy and establishment and maintenance (curation) of ICME databases and libraries (cyberinfrastructures) that will become repositories for this information and that will foster the networking needed to advance a collaborative ICME culture.
- Establishment of policies for promoting broad public access to data and tools generated by federally supported ICME development programs.
- Coordination and monitoring of programs.

The committee notes that many of the materials used by DOD are also important to NASA, so the nation could benefit greatly if DOD were to extend and coordinate its ICME programs with NASA.

National Science Foundation

The committee has concluded that NSF, by supporting ICME-related cross-functional research in its directorates for engineering and mathematical and physical sciences, could transform materials and engineering design. The ICME cyberinfrastructure, described in more detail in Chapter 3, falls within the boundaries of the NSF cyberinfrastructure initiative, so that by supporting the development of the ICME cyberinfrastructure, NSF could also provide an efficient mechanism for sharing the outputs of the fundamental materials research it supports with the engineering community—an essential element for the widespread development of ICME. NSF could also revolutionize ICME database and informatics development by requiring that all data and models whose development it supports be placed in publicly available Web sites that make up the ICME cyberinfrastructure. As discussed in Chapter 2, the National Institutes of Health (NIH) have implemented successful strategies for mandating the public availability of data coming from NIH-funded research. These strategies might prove useful for NSF. Motivating the development of innovative curricula and the training of future professionals in ICME is also an important and essential NSF role.

Recommendation 3: The National Science Foundation—through its Office of Cyberinfrastructure, its Directorate of Engineering, and its Division of Materials Research—should

- **Fund cross-disciplinary research and engineering partnerships to develop the taxonomy, knowledge base, and cyberinfrastructure required for ICME.**
- **Establish incentives and requirements for materials researchers to place their materials information in open-access infrastructures, together with procedures to ensure that the information and models can be used effectively.**
- **Develop engineering talent for ICME by supporting innovative curricula and student internship programs.**

National Institute of Standards and Technology

NIST, as part of the Department of Commerce, has a unique mission: to ensure the competitiveness of U.S. industry. While ICME holds enormous promise for maintaining and enhancing this competitiveness, it will be able to do so only if the formidable data challenges in materials science and engineering are overcome. The committee has concluded that NIST will be able to play a critical role in the development of ICME. No single fixed database exists (nor can one be created) from which all materials engineers can derive the information they need to incorporate materials into a design. On the strength of its mission, however, NIST could establish and curate materials informatics databases and collaborative Web sites, which could then be integrated into the national ICME infrastructure.

Recommendation 4: To promote U.S. innovation and industrial competitiveness, NIST should develop and curate precompetitive materials informatics databases and develop automated tools for updating, integrating, and accessing ICME resources.

Government Support for Small Businesses

As discussed in Chapter 4, small science and engineering companies are playing a key role in developing, advocating, and maturing ICME technologies. They act in effect as scouts for original equipment manufacturers (OEMs), identifying and integrating materials science and engineering (MSE) research into viable commercial products. These innovative firms conduct and capitalize on the kinds of technologies that are required to mature ICME and that the Small Business Innovation Research (SBIR) and Small Business Technology Transfer (STTR) programs were designed to support. In some cases, they may be the ICME experts who provide tool sets to industry.

Recommendation 5: Federal agencies should direct SBIR and STTR funding to support the establishment of ICME-based small businesses.

Coordination of Federal Support for ICME

The committee believes that coordination among federal agencies will be essential to accelerating the progress toward a broad ICME implementation. Many of the committee's recommendations propose similar actions for different agencies—the establishment and curating of accessible databases, for instance. Just as there are many different materials and applications, the committee expects that there will be no single developer or curator of databases or libraries of ICME tools. But to maximize the benefits of ICME to the nation, coordination across the government will be critical. As mentioned earlier, such coordination was recommended by previous National Academies panels but does not appear to have been acted on. The federal government has a mechanism to foster the kind of interagency coordination that would be appropriate for ICME activities. The Networking and Information Technology Research and Development (NITRD)[35] initiative seeks to

- Assure continued U.S. leadership in information technologies to meet federal goals and support twenty-first century government, academic, and industrial interests.
- Accelerate deployment of advanced and experimental information technologies to enhance national and homeland security; maintain world leadership in science, engineering, and mathematics; improve the quality of life; promote long-term economic growth; increase lifelong learning; and protect the environment.
- Advance U.S. productivity and competitiveness through long-term scientific and engineering research in information technology.

Recommendation 6: In pursuit of the promise of ICME to increase U.S. competitiveness and support national security, the Office of Science and Technology Policy should establish an interagency working group under the NITRD to set forth a strategy for ICME interagency coordination, including promoting access to data and tools from federally funded research.

Industry Role

As the primary users of the ICME infrastructure, industry stakeholders must advocate within their organizations for its development and demonstrate early successes to justify continued investment. An important finding of this study is that a return on investment can be realized by following an ICME procedure, even if models are not fully developed or are partially empirical. The key to these successes has been the careful selection of foundational engineering problems, along with mobilizing resources to begin to address these problems and making concerted efforts to introduce ICME in the integrated product development (IPD) process.

With or without major new funding, consortia are an excellent way to organize ICME efforts around collective problems. These consortia could be self-funded by industry or set up as industry-led collectives that approach government agencies for funding, or a combination of both. In one recently initiated consortium in the U.S. auto industry, Chrysler, Ford, and General Motors are developing an ICME infrastructure and knowledge base for magnesium materials and manufacturing processes for auto body applications. This 5-year international program is jointly sponsored by the U.S. Automotive Materials Parnership (USAMP), DOE, China's

---

[35]For more information, see http://www.nitrd.gov/about/about_NITRD.html. Accessed November 2007.

Ministry of Science and Technology, and Natural Resources Canada.[36] The approximate funding over the 5 years is $6 million to $7 million. The program involves participation from researchers at the three auto makers and more than 15 universities and government laboratories.

> **Recommendation 7: U.S. industry should identify high-priority foundational engineering problems that could be addressed by ICME, establish consortia, and secure resources for implementation of ICME into the integrated product development process.**

Role of Academia and Professional Societies

A range of technical and cultural barriers must be surmounted to achieve the vision of ICME. Engineering talent with new skill sets will be needed to develop and use the ICME infrastructure to develop advanced engineering systems. This will require changes in the curricula at universities as well as continuing education for engineers in industry. The University Materials Council (UMC), whose members are the chairs of MSE departments in the United States, is uniquely poised to advocate for the widespread cultural and curricular changes needed to give materials engineers the same computational skills as other engineers and to make ICME a reality. The materials professional societies can also play a key role in removing barriers and accelerating ICME, by organizing conferences and workshops on integrated computational tools in need of development. Materials societies could also serve as a repository for computational materials tools and/or key materials data needed for development and validation of the ICME capability. Another role of professional societies could be to set up continuing education programs that advance the computational skills of their members. The committee recommends as follows:

> **Recommendation 8: The University Materials Council (UMC), with support from materials professional societies and the National Science Foundation, should develop a model for incorporating ICME modules into a broad spectrum of materials science and engineering courses. The effectiveness of these additions to the undergraduate curriculum should be assessed using ABET criteria.**

---

[36]Joseph Carpenter, DOE, "DOE's work on extruded long-fiber-reinforced polymer-matrix composites," Presentation to the committee on March 13, 2007. Available at http://www7.nationalacademies. org/nmab/CICME_Mtg_Presentations.html. Accessed February 2008.

Recommendation 9: Professional materials societies should

- **Foster the development of ICME standards (including a taxonomy) and collaborative networks,**
- **Support ICME-focused programming and publications, and**
- **Provide continuing education in ICME.**

FINAL COMMENT

The committee believes that the MSE discipline is at a critical crossroad and that computationally driven development and manufacturing of materials can be a core activity of materials professionals in the upcoming decades. For the field of materials to keep pace with other engineering disciplines, the development of an ICME infrastructure is essential. Coordination and targeted investment by stakeholders in the critical elements of ICME will allow the following vision to be realized in the next 10-20 years:

- ICME will have become established as a critical element in maintaining the competitiveness of the U.S. manufacturing base.
- ICME practitioners—a broad spectrum of scientists, engineers, and manufacturers—will have open access to a curated ICME cyberinfrastructure, including libraries of databases, tools, and models, thereby enabling the rapid design and optimization of new materials, manufacturing processes, and products.
- The materials scientist in academia performing traditional science-based inquiry will benefit from the assembled and networked data and tools. Discoveries will be easier and their transition to engineering products will be straightforward.
- ICME will have reduced the materials development cycle from today's 10- to 20-year time frame to 2 or 3 years.
- Graduating materials science and engineering students will be employed and operate in a multidisciplinary and computationally rich engineering environment.

2

Case Studies and Lessons Learned

In the 1990s, mechanical engineers began to build and apply integrated computational systems to analyze and design complex engineered systems such as aircraft structures, turbine engines, and automobiles. By integrating structural, fluid, and thermal analysis codes for components, subsystems, and full assemblies, these engineers performed sensitivity analyses, design trade-off studies, and, ultimately, multidisciplinary optimization. These developments provide substantial cost savings and have encouraged the continued development of more advanced and powerful integrated computational systems and tools. Using these systems and tools, aircraft engine design engineers have reduced the engine development cycle from approximately 6 years to less than 2 years while reducing the number of costly engine and subcomponent tests.[1] Integrated product development (IPD) and its primary computational framework, multidisciplinary optimization (MDO), form the core of this systems engineering process (Boxes 2-1 and 2-2).

IPD and MDO have revolutionized U.S. industry, but materials have not been part of this computerized optimization process. While the constraints of diverse materials systems strongly influence product design, they are considered only *outside* the multidisciplinary design loop; an example of this is shown for hypersonic vehicles in Box 2-2. In a large, multidisciplinary engineering environment, an integrated product development team (IPDT)—that is, the group of stakeholders

---

[1]Michael Winter, P&W, "Infrastructure, processes, implementation and utilization of computational tools in the design process," Presentation to the committee on March 13, 2007. Available at http://www7.nationalacademies.org/nmab/CICME_Mtg_Presentations.html. Accessed February 2007.

BOX 2-1
IPD and MDO

A key cultural change introduced into many U.S. industrial sectors toward the end of the twentieth century was the integrated product development (IPD) process, which dramatically improved the execution and efficiency of the product development cycle. IPD is also known as "simultaneous engineering," "concurrent engineering," or "collaborative product development." Its central component is the integrated product development team (IPDT), a group of stakeholders who are given ownership of and responsibility for the product being developed. In an effective IPDT, all members share a definition of success and contribute to that success in different ways. For example, systems engineers are responsible for the big picture. They initially identify development parameters such as specifications, schedule, and resources. During the design process, they ensure integration between tools, between system components, and between design groups. They are also responsible for propagating data throughout the team. Design engineers have responsibilities specific to their capabilities and disciplines. Typically, design engineers determine the scope and approach of analysis, testing, and modeling. They define the computational tools and experiments required to support the development of a design and its validation. Manufacturing engineers ensure that the components can be made with the selected manufacturing process, often defining the computational simulation tools that are required. Materials engineers provide insights into the capabilities and limitations of the selected materials and support development of the manufacturing process.

IPDTs can range in scale from small and focused to multilevel and complex. Regardless of size, however, the defining characteristic of an IPDT is interdependence. The key to a successful IPDT is that the team members, and the tools they use, do not work in isolation but are integrated throughout the design process. This approach may entail communicating outside the original company, country, or discipline.

Owing to the demonstrated success of the IPD process, many engineering organizations, particularly at large companies, have invested considerable human and capital resources to establish a work-flow plan for their engineering practices and product development cycles as executed by the IPDT.

The capability and dynamics of the IPD process are illustrated by the execution of a computationally based multidisciplinary design optimization (MDO). Modern engineering is a process of managing complexity, and MDO is an important computational tool that helps the systems analysts to do that. For example, a modern gas turbine engine has 80,000 separate parts and 5,000 separate part numbers,[1] including 200 major components requiring three-dimensional computer-aided engineering (CAE) analysis with structural finite element and computational fluid dynamics codes. This CAE analysis can easily require over 400 person-years of analytical design and computer-aided design (CAD) support. The only rational way to accomplish this engineering feat organizationally is by means of an IPDT. Owing to the development and validation of computational engineering analysis tools such as finite-element analysis, computer-based MDO has become routine for many systems or subsystems to improve efficiency and arrive at an optimized design or process. Computer-based MDO automates work flow, automates model building and execution, and automates design exploration. A block diagram of the relevant analytical tools utilized by MDO is shown in Figure

---

[1]Michael Winter, P&W, "Infrastructure, processes, implementation and utilization of computational tools in the design process," Presentation to the committee on March 13, 2007. Available at http://www7.nationalacademies.org/nmab/CICME_Mtg_Presentations.html. Accessed February 2007.

2-1-1. Figure 2-1-2 shows the "electronic enterprise" required to support automated MDO. This includes libraries of validated and certified analysis tools; an integration framework; and electronic process or work-flow maps and IPD tools for work-flow management, collaborative engineering, and secure business-to-business information sharing. MDO has allowed the IPDT to focus on product design decisions based on the results of MDO rather than on generating data. This has greatly improved the robustness of final product designs and dramatically reduced the time to reach design solutions.

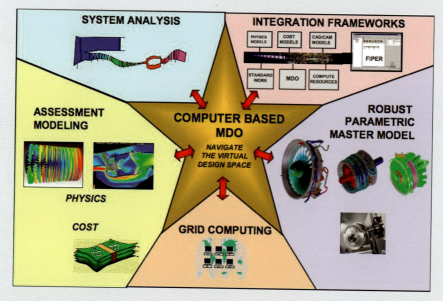

FIGURE 2-1-1 The computational tools required for successful MDO. SOURCE: Michael Winter, P&W, "Infrastructure, processes, implementation and utilization of computational tools in the design process," Presentation to the committee on March 13, 2007. Available at http://www7.nationalacademies.org/nmab/CICME_Mtg_Presentations.html. Accessed February 2007.

who are given ownership of and responsibility for the product under development—operates with materials as a static, limiting constraint on the overall IPD rather than as an optimizable parameter. Typically the list of materials from which a choice is to be made is either taken as a fixed constraint or consists of a small subset of materials that are evaluated statically outside the optimization loop. This approach narrows the design space, resulting in suboptimal vehicle performance in an application with low margins for error. Conversely, the development of the

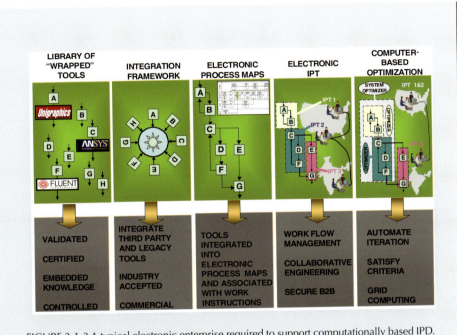

FIGURE 2-1-2 A typical electronic enterprise required to support computationally based IPD. NOTE: Individual codes and tools are depicted as labeled boxes (for example, A is shown as Unigraphics, C is Ansys, F is Fluent, and so on). B2B, business to business. The flow through this figure shows the individual steps involved in establishing the computational process. Adapted from Michael Winter, P&W, "Infrastructure, processes, implementation and utilization of computational tools in the design process," Presentation to the committee on March 13, 2007. Available at http://www7.nationalacademies.org/nmab/CICME_Mtg_Presentations.html. Accessed February 2007.

right class of advanced materials, which is an inherently expensive process, could be better justified and motivated by integration of materials into the MDO computational process.

During the late 1990s, materials engineers who worked on IPDTs witnessed these achievements and began to consider the need for similar computational methods for analysis and development of materials. The need to shrink the growing discrepancy between the system design cycle time and the typical time to develop

BOX 2-2
An Example of MDO

The MDO approach to design has become a powerful vehicle for broad exploration of design space at relatively low cost. An example is the development of hypersonic vehicles for space access and defense applications, where there is a complex interdependence of vehicle structure and aerodynamic and thermal loads; static and dynamic structural deflections; propulsion system performance and operability; and vehicle control. Since many conditions of hypersonic flight cannot reasonably be replicated in any current or foreseeable wind tunnel, designs are not fully validated until actual flights are conducted. Thus the integration of validated analytical design tools with automated data transfer between disciplines in an MDO platform is essential for arriving at realistic but innovative and high-performing solutions. A tool integration scheme used by Boeing Phantom Works for the design of hypersonic vehicles is shown in Figure 2-2-1. This MDO scheme could, for example, be used to design an air-breathing, reusable, hypersonic flight vehicle, where strong interactions between aerodynamics, propulsion, aerothermal loads, structures, and control have a substantial impact on the optimal shape of the entire vehicle. With an MDO platform, thousands of potential vehicle shapes can be explored within the time frame of days (see Figure 2-2-2). Materials tools are notably absent from the MDO tool set (Figure 2-2-1).

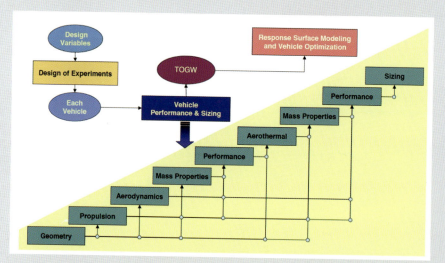

FIGURE 2-2-1 Boeing's integration of analytical tools for design of hypersonic vehicles. SOURCE: K.G. Bowcutt, "A perspective on the future of aerospace vehicle design," American Institute of Aeronautics and Astronautics Paper 2003-6957, December 2003.

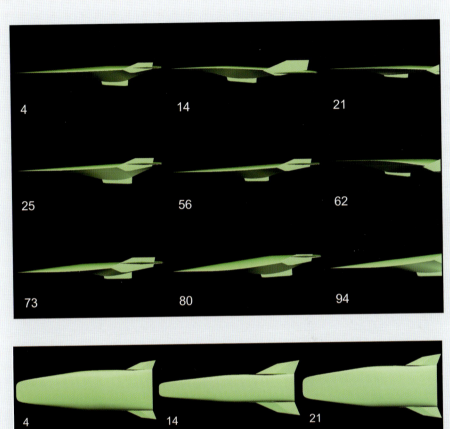

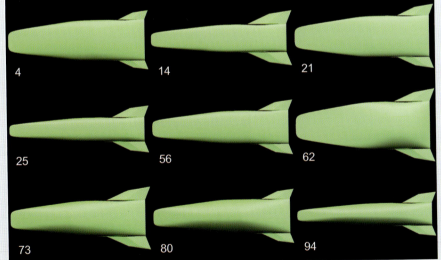

FIGURE 2-2-2 Side and top views of vehicle shape explored in an MDO analysis of hypersonic vehicle design space. This evaluation allows vehicle performance at hypersonic speeds to be explored. A materials analysis is not part of this process. SOURCE: Kevin Bowcutt, Boeing.

a new material (8-20 years) provided further motivation. Materials engineers recognized that by building such tools and methods for materials, a process now termed integrated computational materials engineering (ICME), materials could be incorporated into the overall engineering system, thereby allowing full systems optimization.

Within the last decade, industrial organizations, using their own and government funds, have begun to explore the application of ICME to solve industrial problems, to estimate the payoff for ICME implementation, and to identify development needs. For example, the Defense Advanced Research Projects Agency (DARPA) launched the Accelerated Insertion of Materials (AIM) initiative in 2001 to apply ICME to jet engine turbine disks and composite airframe structures. The goal of AIM was to establish an ICME tool set, integrate materials analysis with design engineering, and demonstrate the benefit of ICME in terms of reduced materials development time and cost. In that same time frame, similar efforts were launched in other industrial sectors to improve product performance and provide information that would be impossible or extremely costly to obtain using experimental methods.[2]

In general, case studies reviewed by the committee demonstrate that while ICME is still an emerging discipline it can nevertheless provide significant benefits, such as lower costs and shorter cycle times for component design and process development, lower manufacturing costs, and improved prognosis of the subsystem component life. In some cases, these benefits would have been impossible to achieve with any other approach, regardless of cost.

In this chapter, several case studies involving ICME are presented along with a discussion of how other scientific disciplines have successfully undertaken large-scale integrated efforts. Case studies where the return on investment (ROI) could be documented are reviewed in more detail. The chapter ends with some lessons learned from these programs.[3]

CASE STUDIES—CURRENT STATUS AND BENEFITS OF ICME

In this section, the current status of ICME is assessed by relating some case studies that illustrate the ICME processes. The vast majority of design engineers see material properties as fixed inputs to their design, not as levers that may be adjusted to help them meet their design criteria. Until materials, component design,

---

[2]John Allison, Mei Li, C. Wolverton, and XuMing Su, "Virtual aluminum castings: An industrial application of ICME," *JOM (Journal of The Minerals, Metals & Materials Society)* 58(11): 28-35.

[3]The committee notes that none of the information provided to the committee was sufficient for the committee to make an independent assessment of the ROI. This is not surprising, because providing such information would involve releasing proprietary or private/classified information. The committee therefore reports the ROI information provided to it without any independent assessment.

and the manufacturing process become fully integrated, designers will not be able to optimize product properties through materials processing. Using an ICME approach, however, this optimization can be accomplished in a virtual environment long before the components are fabricated. The case studies discussed here attempt to show why that is so.

For clarity, the case studies are divided into examples that use ICME to (1) integrate materials, component design, and manufacturing processes; (2) predict component life; and (3) develop manufacturing processes. While each ICME case study achieved different levels of integration and implementation, all of them resulted in substantial benefits. The case studies chosen for inclusion here generally had some level of detailed information on the ROI and other benefits of the ICME activity; this was a self-reported assessment, however, and validating the ROIs was beyond the scope of the committee. The majority of these case studies involved metallic systems. The committee identified similar examples in nonmetallic systems such as polymers and semiconductors, but they either lacked any detailed information on ROI or did not approach full ICME integration. For example, in the area of integrated circuits, models for dielectric constants, electromigration, dislocation generation, and barrier layer adhesion have not yet been integrated with CAD circuit design models or "equipment" models for lithography or processing.[4,5] The committee notes, therefore, that while the examples shown in this report all involve metallic systems, it believes that ICME will be applicable and will demonstrate significant benefits for integrating materials, component design, and manufacturing processes across all materials regimes, including nanomaterials, biological materials, polymeric materials, ceramics, functional materials and composites. However, the challenges can be formidable. For instance, the modeling of polymeric materials is extraordinarily complex at all length scales, and owing to the long chains of the polymer backbones, the gap between feasible atomistic simulations and solid polymer properties is still substantial. Another challenge for industrial polymers is the plethora of suppliers, who offer many proprietary variants of these materials. However, it is this kind of complexity that should motivate an ICME approach.

---

[4]Sadasivan Shankar, Intel, "Computational materials for nanoelectronics," Presentation to the committee on May 29, 2007. Available at http://www7.nationalacademies.org/nmab/CICME_Mtg_Presentations.html. Accessed February 2007.

[5]Dureseti Chidambarrao, IBM, "Computational materials engineering in the semiconductor industry," Presentation to the committee on May 29, 2007. Available at http://www7.nationalacademies.org/nmab/CICME_Mtg_Presentations.html. Accessed February 2007.

Integrating Materials, Component Design, and
Manufacturing Processes in the Automotive Sector

The virtual aluminum castings (VAC) methodology developed by the Ford Motor Company offers one detailed case study for integrating materials, component design, and manufacturing.[6] The methodology was based on a holistic approach to aluminum casting component design; it modified the traditional design process to allow the variation in material properties attributable to the manufacturing process to flow into the mechanical design assessment. Fully funded by Ford to address specific power-train components, the VAC methodology was implemented by the company for cast aluminum power-train component design, manufacturing, and CAE. As discussed below, VAC has resulted in millions of dollars in direct savings and cost avoidance. For VAC to be successful, it required:

- Models of the structure evolution and resulting physical and mechanical properties of aluminum alloy systems, using the classic processing–structure–property flowchart depicted in Figure 2-1,
- The ability to link the various models together while maintaining computational efficiency and simplicity,
- Modifications to the traditional design process to allow for spatial variation of material properties across a component to be considered, and
- Extensive validation of the predictive models.

Quantitative processing–structure–property relationships were defined using a combination of science-based models and empirical relationships and linked to quantitative phase diagram calculations. As shown schematically in Figure 2-1, the influence of all the manufacturing processes (casting, solution treatment, and aging) on a wide variety of critical microstructural features was captured in computational models. These microstructural models were then used to predict the key mechanical and physical properties (fatigue, strength, and thermal growth) required for cast aluminum power-train components.[7] Microstructural modeling was required at many different length scales to capture the critical features required to accurately predict properties. The outputs of this processing–structure–property information are predictions of manufacturing-history-sensitive properties. These predicted properties and their spatial distributions were mapped into the compo-

---

[6]John Allison, Mei Li, C. Wolverton, and XuMing Su, "Virtual aluminum castings: An industrial application of ICME." *JOM (Journal of The Minerals, Metals & Materials Society)*, 58(11): 28-35.

[7]Thermal growth, a common term used for aluminum alloys, represents the dimensional changes brought about by precipitation of phases with different volumes. It is used as a critical input in advanced component durability procedures.

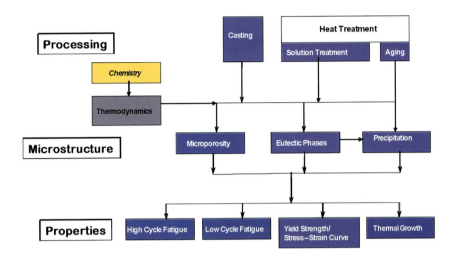

FIGURE 2-1 The processing–structure–property flowchart for cast aluminum alloys for engine components. The chart shows the wide variety of individual models required for ICME. The properties listed are required as the critical inputs to durability (performance) analysis of cast aluminum components.

nent and subsystem finite-element analysis (FEA) of operating engines by which cast aluminum component durability (performance) is predicted.

Both commercial software and Ford-developed codes are used in the VAC system. Commercially available casting-simulation software (Pro-CAST, MagmaSoft), thermodynamic and kinetic modeling software (Pandat, Thermo-Calc, Dictra), first principles code (VASP), and FEA software for stress analysis (ABAQUS) are used, integrating the methodology with the standard codes used in the manufacturing, material science, and design environments. However, many additional codes were developed during the course of the program for specific tasks such as the interfacial heat-transfer coefficient (IHTC) optimization (OptCast), microstructure evolution (MicroMod, NanoPPT, MicroPore), physical or mechanical property evolution (LocalYS, LocalTG, Local FS), and the resultant stresses and component durability (QuenchStress, HotStress, Hotlife). These codes, developed internally by Ford Research and Advanced Engineering, involved coordinating the fundamental research efforts of five universities across the United States and the United Kingdom. Substantial effort was applied to developing efficient links between the output of the casting modeling and structure and property prediction tools to feed seamlessly into the FEA codes and to facilitate reorganization of the design process.

Figure 2-2 shows the process flow for predicting the influence of the casting and heat treatment process on the spatial distribution of yield strength in a typi-

Virtual Aluminum Castings Process Flow

Initial Geometry
•CAD Geometry and Mesh

Filling
•Accurate filling Profile (ProCast, OPTCAST)

Thermal Analysis
•Boundary Conditions (OPTCAST)
•Fraction solid Curves (ThermoCALC)

Microstructure (Al₂Cu)
• Micromodel (MicroMod, PanDat)
• Solution treatment (Dictra)
• Aging Model (NanoPPT, PanDat)

Yield Strength
• LocalYS

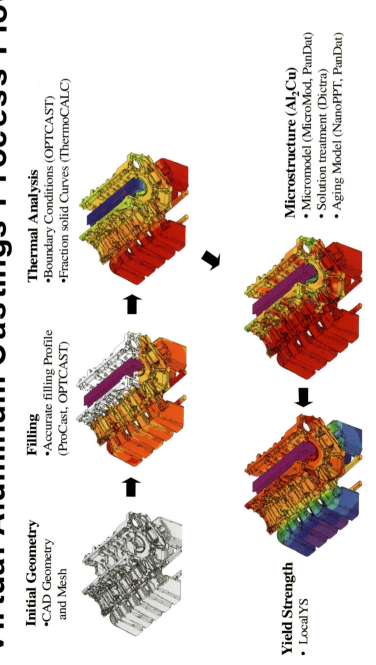

FIGURE 2-2 The ICME process flow for Ford's VAC tool for local property prediction of yield strength (LocalYS). The flow starts with a geometric (CAD) representation of the component. This CAD is then used as input to the simulations of filling and solidification (thermal). The outputs of these simulations are used to predict microstructural quantities. Finally, these microstructural quantities are used to predict the manufacturing history sensitive, spatial distributions in yield strength. Specific commercial codes and Ford-developed programs and subroutines that have been integrated into the process are identified at each step. SOURCE: John Allison, Ford Motor Company.

cal cast aluminum engine block. The finite element or finite difference mesh is the basis for defining the geometry of the part. This geometric mesh is then used with commercial casting software to predict the flow of molten metal during casting and the influence of the casting process and the geometry on the local thermal history. The spatial distributions of the key microstructural features are predicted based on this local thermal history. Finally, the spatial distributions of strength are predicted based on the distribution of these key local microstructural features.

Normally, a manufacturing modeling analysis is initiated only after final design and durability prediction are completed. In the new ICME process, however, the manufacturing analysis is done prior to the durability prediction to (1) provide a more realistic durability prediction and (2) allow for full component design improvements by adjusting both the manufacturing process variables and the component dimensions. Although limited manual optimization of the manufacturing process and product geometry is feasible, computational limitations preclude full multiattribute optimization. Finally, to ensure the acceptance of not only the new tools but also the new design process, extensive validation of the approach was essential. Figure 2-3 shows an example of a typical validation of the LocalYS

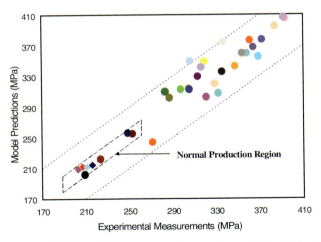

FIGURE 2-3 Validation of ICME predictions are essential to gain acceptance by the engineered product development community. This figure represents a typical experimental validation of VAC local yield strength predictions for a wide range of components, heat treatments, and casting conditions. The data points represent matched pairs of model predictions and experimental validation. The dotted and dashed lines represent statistical bounds. SOURCE: John Allison, Mei Li, C. Wolverton, and XuMing Su, "Virtual aluminum castings: An industrial application of ICME," *JOM* (*Journal of The Minerals, Metals & Materials Society*) 58(11): 28-35.

tool in VAC. Yield strength was measured in a wide variety of locations in castings manufactured under a multitude of processing conditions in a number of different engine components. The necessary degree of accuracy was determined in consultation with product design engineers to build confidence in these predictions. Note the greater model fidelity required in regions of typical production.

In the development of new engines, the cast aluminum engine block and cylinder head are critical components. Before VAC was used, these components were generally developed via iterative engine testing, rework, and retesting. Traditional CAE durability analysis provided a starting point, but due to the lack of influences from manufacturing on properties and residual stresses, this analysis was approximate and these critical components often failed during new engine development. This resulted in costly redesign, retooling, and program delays. The benefits of VAC include a 15-25 percent reduction in new cylinder head or block development time, a reduction in the number of component tests required for assurance testing, and a shorter cycle time for the casting or heat treatment process, giving Ford a cumulative saving of millions of dollars. The VAC system continues to be improved, with future modifications extending it to the high-pressure die-casting process as well as to magnesium alloys, further increasing the potential savings.

The VAC program involved approximately 25 people, according to Ford, and $15 million in expenditures over 5 years. Well over 50 percent of this effort was for experimental work in either model development or validation. The VAC technology is projected to have saved Ford up to $100 million in cost avoidance by resolving product issues and between $2 million and $5 million per year in reduced testing requirements and reduced development time, providing an estimated combined ROI of well over 7:1.[8,9,10] These achievements would not have been possible without both the upfront investment in ICME tool development and the acceptance of a change to the manufacturing culture associated with modifying the design analysis to allow the results of manufacturing simulations of the component to be used as inputs to the component life predictions. VAC was initially developed for a single alloy system. It served as a platform for the ICME development of additional cast aluminum alloys at a substantially lower cost. For example, a recently developed VAC tool set for a new alloy for aluminum cylinder heads for diesel engines was completed in less than half the time and at a cost of between 10 and 20 percent of that required for development of the initial VAC tool set. Ford has extended the ICME approach to coupling analysis of sheet steel stamping with predicting

[8]John Allison, Mei Li, C. Wolverton, and XuMing Su, "Virtual aluminum castings: An industrial application of ICME," *JOM (Journal of The Minerals, Metals and Materials Society)* 58(11): 28-35.

[9]Private communication, John Allison, July 2007.

[10]NRC, *Retooling Manufacturing: Bridging Design, Materials, and Production*, Washington, D.C.: The National Academies Press (2004).

crash performance for vehicle bodies. It is also developing advanced approaches for coupling the casting analysis of magnesium and aluminum body components with crash performance predictions.

Integrating Materials, Component Design, and Manufacturing Processes in the Aerospace Sector

The DARPA AIM program is another example of an effort to integrate material properties in the design and manufacturing optimization process. As depicted in Figure 2-4, when the program was initiated in 2001, the materials discipline was not yet integrated into the turbine engine design stream. Within 1 year, material behavior models were integrated sufficiently to enable the execution of statistically designed matrices and generation of response surfaces. Within 2 years, the material behavior models were fully integrated, enabling execution of the several case studies (described below). At that point the fundamental infrastructure was in place,

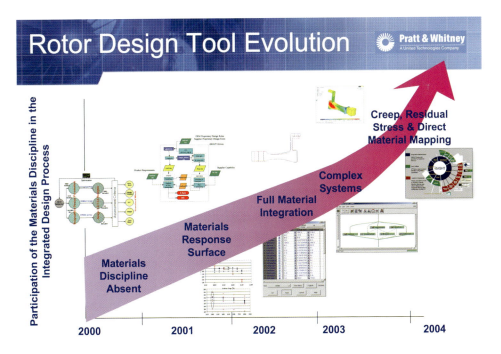

FIGURE 2-4 Integration of the materials discipline into the aeroengine disk design process at P&W during the DARPA AIM program. Over the 4 years, the materials discipline was at first absent altogether from the automated design process. It then participated as statistically derived response surfaces and advanced to integration as physically based models, eventually becoming integrated into complex mechanical systems with multiple material performance characteristics.

facilitating the addition of incremental capabilities (complex systems, additional material properties). Once the significant investment in the underlying ICME infrastructure had been made, the incremental investment required to extend the capability and applicability was significantly reduced.

In its AIM program, Pratt & Whitney (P&W) used this approach to optimize the design and manufacturing of jet engine turbine rotors. The design figures of merit were rotor weight and burst speed. Decreasing rotor weight provides both direct cost savings and improved system performance. Increasing rotor burst speed improves system performance. The maximum benefit in both figures of merit was achieved with the simultaneous optimization of the disk design (geometry) and the manufacturing process (material properties), requiring the integration of commercially available engineering analysis codes and material behavior models developed at universities. An initial constraint in this program was the lack of appropriate mechanical property models for the turbine rotor materials. Once the models were developed, it became apparent that key features of material structure (precipitate sizes and distributions, for example) were not available for use in the models developed. Thus, in spite of the fact that the material had been in service for many years, substantial materials characterization was still required.

A key element of the analytical code integration was the use of optimization software in conjunction with a parametrically defined geometry in the FEA codes, enabling both the component configuration and material processing to be con-currently optimized. The ICME process was validated in three different exercises to determine how well the methodology would improve the burst capability of a turbine engine rotor, as illustrated in Figure 2-5.

Taking only the rotor weight reductions into consideration, system life-cycle benefits on the order of $200 million have been projected to be possible for a typical turbine engine application and fleet. Reductions of $3 million in the cost of design data testing are achievable with a sixfold reduction in test costs. To fully realize these benefits, the ICME methodology needs to be expanded to include other critical models of material behavior. This would require an additional $20 million investment. Taking into consideration the increased investment, these estimated benefits have been projected to yield an ROI of around 10:1 for a fully developed ICME methodology.[11] However, the investments needed to extend this approach in a more pervasive fashion to other material systems used by P&W have not been forthcoming. Chapter 4 discusses industrial inertia in some detail.

---

[11]Private communication with Jack Schirra, committee member; also, Leo Christodolou, DARPA, "Accelerated insertion of materials," Presentation to the committee on November, 20, 2006. Available at http://www7.nationalacademies.org/nmab/CICME_Mtg_Presentations.html. Accessed February 2007.

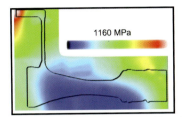

Yield Strength Distribution in Forging
(Disk Geometry Superimposed)

Disk Burst Test

Case Study	Heat Treatment	Forging Parameters	Component Geometry	Forging Weight	Disk Burst Speed	Comments
1	Constant	Variable	Variable	-18%	+6%	Current state of art
2	Variable	Variable	Constant	-11%	+12%	Final shape constrained
3	Variable	Variable	Variable	-21%	+19%	Full impact of integration

	Cost Benefit	System Benefit

FIGURE 2-5 Full integration of materials science with design process: turbine disk validation studies. This chart summarizes the application of integrated engineering analysis tools to a series of case studies for aeroengine disk applications. Models integrated: component processing, microstructure evolution, microstructure-based strength model, component geometry, and stress. Case Study 1 represents the current state of the art and shows that integration of structural engineering analysis tools with part design can identify optimal configurations that result in measurable performance improvements. Case Study 2 represents the scenario where ICME enables the improvement of a production part capability without requiring a system-level redesign and product reconfiguration. The cost for a rotor redesign and reconfiguration effort can be on the order of $200 million depending on fleet size. Case Study 3 represents the full application of ICME in the design process and shows that system-level performance can be substantially improved over the current state of the art.

Integrating Materials, Component Design, and Manufacturing Processes: Other Examples

Other examples of integrating materials, component design, and manufacturing processes can be found in Table 2-1, along with their corresponding benefits. In all of these ICME efforts, the potential cost savings are significant. It is worth noting, however, that in many cases maximizing benefits from ICME will require additional investments and may encounter the cultural barriers identified in Chapter 4.

While some ICME efforts represent large programs at major manufacturers, Table 2-1 shows that small companies are also making important contributions

TABLE 2-1 Examples of ICME for Integrated Manufacturing, Materials, and Component Design

Company	Case Study	Benefits
Ford Motor Co.	Virtual aluminum castings	15-25 percent reduction in product development time, substantial reduction in product development costs, optimized products (greater than 7:1 ROI)
General Electric Company/ P&W/Boeing	Accelerated Insertion of Materials	Potential 50 percent reduction in development time, potential eightfold reduction in testing, potential improvement in component capability
Livermore Software Technology Corporation/ESI Group/Ford (separate studies)	Use of stamping CAE output as material property input for crash CAE in automotive structures	Enabled use of advanced high-strength steels in body structures for significant weight savings
Naval Surface Warfare Center	Development of an Accelerated Insertion of Materials system for an aluminum extrusion	Development in progress; optimization of thermomechanical history of AA6082 during the extrusion of sidewall panels for a littoral combatant ship
Knolls Atomic Power Laboratory (Lockheed Martin Corporation)/Materials Design	Fracture of high-strength superalloys in nuclear industry	Optimization of critical embrittling/ strengthening impurities
Toyota central R&D labs/ Materials Design	Clean Surfaces Technology Program—develop dopants for UV photocatalyst	Reduced time to identify new dopants, new products (air purifiers)
QuesTek	Alloy development	Reduced insertion risk, reduced experimental cost
Boeing	Airframe design and manufacturing	Development in progress, Boeing estimates a reduction in material certification time by 20-25 percent (3-4 years)

to ICME, often with government sponsorship. A number of small companies that offer new tools for developing new alloys or processes are in the start-up stage, and are somewhat reliant on government funding. QuesTek has developed several alloys for both commercial (automotive, consumer products) and government (Army, Navy, NASA) applications, while Materials Design, Inc., has developed materials varying from ultraviolet catalysts (Toyota) to optimized superalloy chemistries for the nuclear industry (Knolls Atomic Power Laboratory). The committee concluded that the agility of small companies is an asset in developing new approaches to the design process, including ICME.

Integrated Materials Prognosis (Component Life Estimation)

In some cases, quantifying a component's lifetime is irrelevant; the component will last longer than the system. However, for certain components, the lifetime is critical both for design decisions and for subsequent system maintenance and support. The current methodology for predicting lifetime is based on accelerated aging tests. Such tests are expensive, uncertain, and sometimes impossible. By integrating component environment with material property evolution, ICME can be used to predict bounds on component lifetime. The longer the lifetime of the component, the greater the savings associated with such an ICME approach.

One study involving an integrated materials prognosis used synergistic computational and experimental efforts at Los Alamos National Laboratory (LANL) and Lawrence Livermore National Laboratory (LLNL). In the nuclear weapons stockpile, credible estimates of component lifetimes are required to plan for the future refurbishment and manufacturing needs of the weapons complex. One of the most important of these components is the pit, that portion of the weapon that contains the fissile element plutonium. The U.S. government had proposed construction of a modern pit facility, and a key variable in planning both the size and schedule for this facility is the minimum estimated lifetime for stockpile pits. To support this effort, the National Nuclear Security Administration (NNSA) of the DOE sponsored a program to provide predictions of primary-stage pit lifetimes owing to plutonium aging. Because accelerated aging tests of plutonium are difficult and expensive, and performance tests of pits are prohibited by national policy (in voluntary compliance with international treaty), an integrated computational approach was required. The funded program was based on analyses of archival underground nuclear-explosion testing (UGT) data, laboratory experiments, and computer simulations. By combining (1) the results of past UGTs with pits of various ages, (2) experimental and theoretical investigations of the metallurgical properties of plutonium containing various combinations of impurities, and, finally, (3) computer simulations of primary performance with model plutonium properties varying with age, the groups at LANL and LLNL were able to provide estimates of pit lifetimes such that any new pit facility construction could be scheduled appropriately.[12]

A combination of commercial codes (Thermo-Calc, QMU-Crystal Ball), molecular dynamics codes, design sensitivity tools, and specific codes developed

---

[12]Dimitri Kusnezov, National Nuclear Security Administration, "Mission challenges in large-scale computational science," Presentation to the committee on December, 1, 2006. Available at http://www7.nationalacademies.org/nmab/CICME_Mtg_Presentations.html. Accessed February 2007.

for this particular program was used. The laboratories used the Quantification of Margins and Uncertainty (QMU) methodology to assess pit lifetimes based on simulations of primary performance. Age-dependent models derived from data, science-based computational methods, and conservative assumptions were used to calculate the change in pit lifetime owing to the change in material properties. To quote a review of this program, "For these systems it is important that each contribution to the lifetime be well understood and validated. In a sense, the issue is not one of accounting for aging but of managing the margins and uncertainties that are already present at zero age, and this is best done by understanding the trade-offs involved and the consequent mitigation strategies that can be applied."[13]

The LANL/LLNL pit lifetime prognosis program involved approximately 200 people over 7 years, at a total cost of $150 million—clearly a large-scale effort. Because the results showed much longer pit lifetimes than initially expected, the construction of a manufacturing facility with an approximate cost of $1.5 billion was not required, giving an ROI of 10:1. Even if a manufacturing facility had been required, this ICME program cost no more than an equivalent accelerated aging effort, and provided more reliable results.[14]

One key lesson from this case study was that only 10-20 percent of the program cost was for computation. A heavy investment was made in experimental data to provide subscale/accelerated validation for the models. The cost of computation for this application was modest because the program leveraged the prior investments of DOE's Advanced Scientific Computing program.[15,16] So the reported $150 million cost considerably underestimates the actual cost. The fact remains, however, that not only was the ROI quite high, but these results could not have been provided using conventional prototyping and component testing without violating national nuclear policies.

Integrated materials prognosis is perhaps the least mature of any of these areas of ICME. However, there are a few other examples of current programs that

---

[13]R.J. Hemley, D. Meiron, L. Bildsten, J. Cornwall, F. Dyson, S. Drell, D. Eardley, D. Hammer, R. Jeanloz, J. Katz, M. Ruderman, R. Schwitters, and J. Sullivan, *Pit Lifetime,* National Nuclear Security Report JSR-06-335, U.S. Department of Energy, January 11, 2007, p. 17.

[14]Louis J. Terminello, LLNL, "Synergistic computational/experimental efforts supporting stockpile stewardship," Presentation to the committee on March 13, 2007. Available at http://www7.national academies.org/nmab/CICME_Mtg_Presentations.html. Accessed February 2007.

[15]Joseph C. Martz, and Adam J. Schwartz. "Plutonium: Aging mechanisms and weapon pit lifetime assessment," *JOM,* September 2003.

[16]R. J. Hemley, D. Meiron, L. Bildsten, J. Cornwall, F. Dyson, S. Drell, D. Eardley, D. Hammer, R. Jeanloz, J. Katz, M. Ruderman, R. Schwitters, and J. Sullivan, *Pit Lifetime,* National Nuclear Security Report JSR-06-335, U.S. Department of Energy, January 11, 2007, p. 17.

TABLE 2-2 Examples of ICME for Integrated Material Prognosis (Life Estimation)

Company or Organization	Case Study	Benefits
IBM/Intel	Electromigration	Not provided
P&W/General Electric	Engine systems prognosis	In progress; potential to provide substantial savings for maintenance of military fleet
Lockheed Martin/Sandia National Laboratories/ U.S. Navy	Aging of solder joints	No other feasible way to estimate 50-year life expectancy

are ongoing, and the benefits of these programs are still only estimates of future implementation, as shown in Table 2-2.

Developing a Manufacturing Process

The most widespread applications of ICME have been in the component manufacturing process. While large-deformation finite element analysis (FEA) and computational fluid dynamics (CFD) commercial codes are now widely used for process modeling, the incorporation of microstructure evolution in these simulations is relatively new, indeed in its infancy in terms of commercialization. For the early adopters, there exists the potential for improving yields, reducing cycle time and energy costs, and optimizing a process for a given material or desired structure.

The Controlled ThermoMechanical Processing (CTMP) of tubes and pipes program provides one case study of the integration of materials modeling into the development of a manufacturing process. The Timken Company makes steel tubes that are subsequently formed into bearings, gears, or pipes. Tube fabrication is a batch process, and customers request alloys of diverse composition and properties. Historically, both forming and annealing processes were specified in an ad hoc manner, which decreased yields. The ultimate goals of the CTMP program were reduced process variability as well as optimizing the tube-making process by means of real-time control. Improved control and more consistent properties lead to increased yields, improved performance in customer applications, and reduced energy, cost, and cycle time. This program, led by the Alloy Steel Business of the Timken Company, was co-funded by DOE, with Timken providing approximately 50 percent of the cost. Program costs totaled approximately $10 million over 5 years and required extensive collaboration among more than 25 different industrial, academic, and government facilities in the United States and around the world.

The main components of the CTMP program were the development of analytical tools, measurement techniques, process simulation, and product response. The two most important analytical tools of the program were the tube optimization model (TOM) and the virtual pilot plant (VPP). TOM offers a PC-based framework for developing a tube-making process that generates the desired microstructure. VPP allows process and equipment scenarios to be evaluated in computer simulations rather than in the production facilities. The TOM was initially developed for use with three different grades of steel; however, many grades were added to make the models more broadly applicable. Timken made extensive use of a "design of experiments" approach to develop data-driven relationships between controllable factors in the manufacturing process and the material microstructure, which were then hard-wired into the TOM.[17] The majority of codes in TOM were developed within the program, such as those that describe controlled slow cooling, inverse heat conduction, austenite decomposition, thermal expansion coefficients, recrystallization, grain growth, flow stress, and some specialized finite element mill models (ELROLL). Commercial codes used were QuesTek MCASIS codes for continuous cooling transition curves, various finite element (FE) or finite difference (FD) codes (ABAQUS, DEFORM, SHAPE), and optimization codes such as EPOGY in the VPP. Direct, on-line measurements of austenitic grain size using a laser-ultrasonic gauge and an eccentricity gauge were used to validate in-process grain size, building on an earlier DOE program in this area. TOM was used to assess the impact of various manufacturing process parameters on the machinability of the finished material, a critical customer requirement. For the case of machining tubes into automotive gears, this capability allows the steel microstructures to be optimized to extend the lifetime of broaching tools.

The benefits of the CTMP program can be demonstrated by reviewing some of its elements. The thermal-enhanced spheroidization annealing (T-ESA) "process recipe" is an annealing cycle that reduces the time and energy requirements involved in the spheroidization heat treatments applied to 52100 and other homogeneous high-carbon steels. The recipe has been perfected and implemented, reducing the process cycle time 33-50 percent and energy costs by $500,000 per year. The accumulated annual savings of other direct benefits are estimated at nearly $1 million, not to mention the opportunity cost that the tools provided by reducing the time that would be required of the technical staff to execute those studies without TOM. Moreover, TOM mill simulation and a mill trial demonstrated, with limited experiments, a strong potential for application of a minimum-capital manufacturing process. If new capital equipment were installed, the benefits from a single

---

[17]Design of experiments is a systematic approach to investigation of a system or process. A series of structured tests is designed in which planned changes are made to the input variables of a process or system. The effects of these changes on a predefined output are then assessed.

application of TOM/VPP would be in the millions of dollars. Finally, a preferred microstructure was discovered for maximizing gear broach tool life. Since cutting tools cost roughly half as much as gears, the industry-wide savings could amount to more than $10 million annually.

Obviously not all ICME case studies have realized their full potential. For example, while the direct financial benefits at Timken were reportedly limited because there was no capital investment in an optimized plant, the experience there shows it is reasonable to expect that an ROI of 3 to 10 could be realized over the entire industry if capital investments were made. Commercialization of some of the codes developed in the CTMP program is planned, however, and would increase the use of ICME in this industry. The CTMP program has demonstrated that science developed into applicable technology can differentiate domestic products and further the cause of increasing U.S. competitiveness in the global market.[18]

There are a number of other examples in which manufacturing vendors have begun to embrace ICME, as shown in Table 2-3. While this area of integrated materials and manufacturing promises to provide the most immediate benefits to industry, it must be recognized that additional development and validation remains for these new techniques to be more widely utilized.

LESSONS LEARNED FROM OTHER DISCIPLINES

In addition to investigating the current status of ICME, the committee also explored some major integration efforts in other scientific and engineering fields. Integration experiences in other disciplines teach important lessons about accomplishing major technical and cultural shifts. It is important to note that fields that have achieved successful integration share some advantages over materials engineering. First they enjoy a cohesive data structure, a common mathematical framework, and well-defined objects for investigation. For example, genomics describes a single kind of data, gene sequences, while astronomy focuses on a common set of celestial objects. In contrast, materials engineers working on a large array of engineering components must know about the physical and mechanical properties of the materials used as well as their spectroscopic and two-dimensional and three-dimensional microscopic characteristics. Given the diversity of the information and how this information is applied, the materials challenges more closely resemble the challenges of bioinformatics. In general, communities that have benefited from integration of information have set explicit goals for information acquisition and

---

[18]For more information on the CTMP program, see Timkin Company, *Final Report: Controlled Thermo-Mechanical Processing of Tubes and Pipes for Enhanced Manufacturing and Performance*, November 11, 2005. Available at http://www.osti.gov/bridge/purl.cover.jsp?purl=/861638-qr9nuA/. Accessed May 2008.

TABLE 2-3 Examples of ICME in Materials Manufacturing Processes

Company	Case Study	Benefits
The Timken Company	CTMP	Potential to reduce manufacturing cost by 20 percent with new heat treat/mill on-line
Ladish Co./Rolls-Royce/General Electric Company/P&W/Boeing/Timet	Titanium modeling	The summary of the business case for this project is a 4.83:1 ROI with a $24.442 million ROI within a standard 10-yr analysis period
Ladish Co./General Electric Company	Processing science methodology for metallic structures	Cost avoidance for production of forged aerospace components
Special Metals Corporation (a PCC company)	Four case studies in alloy development using Thermo-Calc	Savings of more than $500,00 in development materials, estimated $1.5 million in new revenue
Böhler Schmiedetechnik	Optimizing the forging of critical aircraft parts by the use of finite element coupled microstructure modeling	Cost avoidance for production of forged aerospace components
Carmel Forge/Scientific Forming Technologies Corporation	Grain size modeling for Waspalloy	Cost avoidance for production of forged aerospace components
RMI Titanium Company	Rolling modeling for development of fine grain titanium sheet at RMI	Faster, more cost-effective optimization of rolling parameters and pass schedules
Alcoa Howmet	Grain size modeling	Improved prior beta grain size control in investment casting of Ti-64

successfully executed "team science" projects that have established national centers and open-access databases of fundamental information. In the following sections specific lessons learned from the genomics, bioinformatics, and astronomy communities are highlighted.

Genomics and Bioinformatics

One of the most significant integration efforts in modern science was the human genome project (HGP).[19] This well-coordinated, $3 billion, 13-year project

---

[19]For more information, see http://www.ornl.gov/sci/techresources/Human_Genome/home.shtml. Accessed October 2007.

was funded in the United States by the DOE and the National Institutes of Health (NIH) and was conducted in collaboration with the United Kingdom, Japan, France, Germany, China, and other countries. The project determined the complete sequence of the three billion DNA subunits (bases), identified all human genes, and made them accessible for further biological study. At any given time, HGP involved over 200 researchers, and its successful completion required large-scale funding, the coordination of important technologies (for example, rapid, high-throughput sequencing capabilities and databases), and the evolving principles surrounding intellectual property and publication.[20] All human genomic sequence information generated by the centers that had been funded for large-scale human sequencing was made freely available in the public domain to encourage research and development and to maximize the benefit to society. Further, the sequences were to be released as soon as possible and finished sequences submitted immediately to public databases. To promote coordination of activities, it was agreed that large-scale sequencing centers would inform the Human Genome Organization (HUGO) of their intention to sequence particular regions of the genome. The information was presented on the HUGO Internet page and directed users to the Web pages of individual centers for more detailed information on the status of sequencing in specific regions. Although HGP was completed in 2003, NIH continues to fund major coordinated sequencing projects. As an example of the magnitude and type of efforts funded by NIH, the sequence of the Rhesus monkey was recently completed, the result of a $20 million effort involving 100 researchers that was approved in 2005.[21]

While the resources expected to be made available to ICME are likely to be fewer than were applied to mapping the human genome, there are significant lessons to be learned by the ICME community from HGP by considering how that community initially organized and set goals and by considering the potential impact of large-scale, coordinated projects based on the gathering and organization of data. At the outset of the HGP, quantitative goals for information acquisition (gene sequencing), databases, computational advances, and the human infrastructure were set. In 1991, 5-year goals were established, with funding of $135 million from NIH and DOE. The 1991 goals included these:[22]

- Improve current methods and/or develop new methods for DNA sequencing that will allow large-scale sequencing of DNA at a cost of $0.50 per base pair.

---

[20]Rex Chisholm, Northwestern University, "Community computational resources in genomics research: Lessons for research," Presentation to the committee on May 29, 2007. Available at http://www7.nationalacademies.org/nmab/CICME_Mtg_Presentations.html. Accessed February 2007.

[21]Elizabeth Pennisi, "Boom time for monkey research," *Science* 316 (April): 216.

[22]For more information on the HGP's 5-year plan, see http://www.genome.gov/10001477. Accessed February 2008.

- Develop effective software and database designs to support large-scale mapping and sequencing projects.
- Create database tools that provide easy access to up-to-date physical mapping, genetic mapping, chromosome mapping, and sequencing information and allow ready comparison of the data in these several data sets.
- Develop algorithms and analytical tools that can be used in the interpretation of genomic information.
- Support research training of pre- and postdoctoral fellows starting in FY 1990. Increase the numbers of trainees supported until steady-state "production" of about 600 per year is reached by the fifth year.

The technology for rapid, low-cost sequencing advanced quickly, while the needs for databases and computational tools continued to evolve.[23] Interestingly, the training goals were not met, "because the capacity to train so many individuals in interdisciplinary sciences did not exist." The establishment of interdisciplinary research centers, with significant participation from nonbiological scientists, was identified as an ongoing need throughout the program.[24] ICME shares many of the same challenges: rapid, low-cost experimentation, development of databases and computational tools, and the need for interdisciplinary training.

The HGP also drove developments in bioinformatics, a field that lies at the intersection of biology, computer science, and information science and is defined by the NIH as "research, development, or application of computational tools and approaches for expanding the use of biological, medical, behavioral or health data, including those to acquire, store, organize, archive, analyze, or visualize such data." Bioinformatics generally makes use of publicly available databases that can be mined for associating complex disorders (analogous to material properties) with different versions of the same gene (analogous to microstructures). In the case of NIH-sponsored genetic research, genetic data must be published in publicly accessible databases. Similarly genetics-oriented research journals also require that this kind of information is made publicly available before the relevant paper may be published. The length scales associated with bioinformatics data and their application pose a challenge as complicated as materials. Examples of informatics databases include GenBank (genetic sequences), EMBL (nucleotide sequences), SwissProt (protein sequences), EC-ENZYME (enzyme database), RCSB PDB (three-dimensional biological macromolecular structure data from X-ray crystallography), GDB (human genome), OMIM (Mendelian inheritance in man

[23]F. Collins and D.J. Galas, "A new five-year plan for the U.S. Human Genome Project," *Science* 262 (1993): 43-46.

[24]Francis S. Collins, Ari Patrinos, Elke Jordan, Aravinda Chakravarti, Raymond Gesteland, and LeRoy Walters, "New goals for the U.S. Human Genome Project: 1998-2003," *Science* 282 (1998):682-689.

data bank), and PIR (protein information resource). The National Center for Biotechnology Information (NCBI),[25] which provides access to these and other databases, was created in 1998 as a national resource for molecular biology information. The mission of the NCBI includes the generation of public databases, the research and development of computational biology tools, and the dissemination of biomedical information.

An important lesson learned by the genetics community is that a key first step in establishing standards is agreement on a taxonomy (that is, an agreed-on classification scheme) and a vocabulary that ensures interoperability of data and models.[26] Database curators play a key role in ensuring the quality of the data and determining what data are needed by the community. Bioinformatics database development, maintenance, and curation are funded by NIH. A small model organism database might cost $400,000 annually, including the services of three curators, two database programmers, and the principal investigator.[27]

One of the cultural lessons from bioinformatics is that transitioning from bench science to big science involves a recognition on the part of the researchers that although their data are connected to them in the database, once those data are used by others they are no longer connected to them. This transition also requires a realization that analysis of a researcher's data by others does not diminish them but increases their impact. One presentation suggested that this required a transformation from a "hunter-gatherer" research model to a "collective farming" model, in which coordination and collaboration are the central elements.[28]

There are still no similar, publicly funded databases, informatics efforts, or comprehensive training programs for ICME. In fact, there are substantial barriers to developing them: lack of funding, lack of standards, diverse data classes, proprietary data ownership, and cultural barriers to building the needed collaborations among materials science, engineering, computational science, and information technology. However, there is a clear opportunity to capitalize on the large body of public high-quality data by establishing open-access, mineable databases analogous to those in the bioinformatics world. An effort on the part of the emerging ICME community to set quantitative and specific goals for materials characterization, databases, computational models, and training will be needed for this discipline to mature.

---

[25]For more information see http://www.ncbi.nlm.nih.gov/. Accessed February 2008.

[26]Rex Chisholm, Northwestern University, "Community computational resources in genomics research: Lessons for research," Presentation to the committee on May 29, 2007. Available at http://www7.nationalacademies.org/nmab/CICME_Mtg_Presentations.html. Accessed February 2008.

[27]Cate L. Brinson, Northwestern University, "Materials informatics—what, how and why: Analogy to bioinformatics," Presentation to the committee on May 30, 2007. Available at http://www7.nationalacademies.org/nmab/CICME_Mtg_Presentations.html. Accessed February 2008.

[28]Ibid.

Open Science Grid and Sloan Digital Sky Survey

Fields from climate change to gaming have also benefited from integration of scientific models and collaborative frameworks involving large-scale databases or computing needs. Some efforts such as climate change modeling and genetics involve large-scale, central coordination; others such as gaming are more organic in nature and based on a free, open source software (FOSS) paradigm.[29] These activities have been enabled largely by the Internet and tend to be Web based, with networks of geographically distributed scientists and code developers working together. This process, known in the United States as cyberinfrastructure and in Europe as e-Science, has given rise to the new field of grid computing, in which computers are shared.[30]

The committee was briefed on two such efforts, the Open Science Grid (OSG)[31] and the Sloan Digital Sky Survey.[32,33] These efforts were initiated because of the need for large-scale (petascale) computing and storage within the astronomy and physics community. The Sloan Digital Sky Survey provides over 40 terabytes (TB) of raw data and 5 TB of process catalogs to the public. The data challenge in this field is the integration of disparate types of data about astronomical objects (stars, galaxies, quasars), including images, spectroscopy data (acquired by an array of experimental techniques at various wavelengths), and astrometric data, along with the large volumes of data (2 to 4 TB per year). Tools developed for the automated data reduction efforts that make the survey possible have involved more than 150 person-years of effort. A lesson learned in this activity was that information is growing exponentially and planning for this data explosion is important.

The OSG comprises a grid or distributed network of over 70 sites on four continents accessing more than 24,000 central processing units. Establishment of

---

[29]W. Scacchi, "Free and open source development practices in the game community," *IEEE Software* (January 2004): 59-66.

[30]Daniel Clery, "Infrastructure: Can grid computing help us work together?" *Science* 313 (July 2006): 433-434.

[31]OSG is a consortium of software, service, and resource providers and researchers from universities, national laboratories, and computing centers across the United States. It brings together computing and storage resources from campuses and research communities into a common, shared grid infrastructure over research networks via a common set of middleware. The OSG Web site says the grid offers participating research communities low-threshold access to more resources than they could afford individually. For more information on the OSG, see http://www.opensciencegrid.org/. Accessed February 2008.

[32]Paul Avery, University of Florida, "Open Science Grid: Linking universities and laboratories in national cyberinfrastructure," Presentation to the committee on March 13, 2007. Available at http://www7.nationalacademies.org/nmab/CICME_Mtg_Presentations.html. Accessed February 2008.

[33]Alex Szalay, Johns Hopkins University, "Science in an exponential world," Presentation to the committee on May 30, 2007. Available at http://www7.nationalacademies.org/nmab/CICME_Mtg_Presentations.html. Accessed February 2008.

this grid was funded (1999-2007) by $38 million from DOE and NSF. The OSG serves a diverse set of disciplines including astronomy, astrophysics, genetics, gravity, relativity, particle physics, mathematics, nuclear physics, and computer science. According to information provided to the committee, a key lesson learned from the OSG is that over half the challenges associated with its establishment were cultural. Thus significant effort was and is required in project and computational coordination and management, education, and communication. The OSG has a well-developed communication Web site, monthly newsletters, and annual summer schools for participants. Key technical challenges include commercial tools that fall short of the needs for grid computation, requiring OSG collaborators to invent the software.

SUMMARY AND LESSONS LEARNED

Several consistent lessons emerged from the case studies reported to the committee. These lessons set out a context for a path forward for ICME, which is discussed in Chapter 3 and Chapter 4.

Lesson Learned 1. ICME is an emerging discipline, still in its infancy.

Although some ICME successes have been realized and articulated in the case studies, from an industrial perspective ICME is not mature and is contributing only peripherally. While some companies may have ICME efforts, product design often goes on without materials modeling. Significant government funds have been expended on developing tools for computational materials science (CMS), and many of these tools are sufficiently advanced that they could be incorporated into ICME models. However, government and industry efforts in integrating CMS tools and in applying them to engineering problem solving are still relatively rare.

Lesson Learned 2. There is clearly a positive return on investment in ICME.

Performance, cost, and schedule benefits drive the increasing use of simulation. ICME shows promise for decreasing component design and process development costs and cycle time, lowering manufacturing costs, improving material life-cycle prognosis, and ultimately allowing for agile response to changing market demands. Typical reductions in product development time attributed to use of ICME are estimated to be 15 to 25 percent, with best-case ROIs between 7:1 and 10:1. Less quantifiable, but potentially more important, ICME often offers solutions, whether for design decisions or lifetime prognoses, that could not be obtained in any other way.

Lesson Learned 3. Achieving the full potential of ICME requires sustained investment.

Because most ICME efforts are not yet mature, they are just beginning to realize benefits from the investment in them. Several case studies, including the CTMP and AIM programs, highlighted the need for additional, sustained commercial investments, whether in human resources, code development, or capital equipment, to achieve the full ICME payoff. Government programs, in particular, often fund the initial investigation of a concept but leave follow-up to others. Developing ICME into a mature discipline will take a considerable investment from government and industry.

Lesson Learned 4. ICME requires a cultural shift.

Several case studies articulated that the cultural changes required to fully benefit from ICME should not be underestimated. For ICME to gain widespread acceptance, shifts are required in the cultures in industry, academia, and government. The design philosophy must shift from separating the product design analysis and the manufacturing process optimization. The engineering culture must shift toward increasing confidence in and reliance on computational materials engineering models and depending less on databases and physical prototypes. Materials researchers and other data generators must shift toward an open-access model that uses data in a standard format. Each of these cultural changes is challenging and must be supported with education and resources.

Lesson Learned 5. Successful model integration involves distilling information at each scale.

Models that explicitly link different length scales and timescales are widely viewed as a laudable goal; however, the very few cases where this goal has been met required computational resources that are not widely available. In most successful ICME case studies, length scales and timescales were integrated by reducing the information at each scale to simple, computationally efficient models that could be embedded into models at other scales. By focusing on ICME as an engineering undertaking, this approach to incorporating information on the length scale, the timescale, or the location was found to be effective. However, it requires experts with sufficient understanding of a particular materials system to be able to judge which material response issues are essential.

Lesson Learned 6. Experiments are key to the success of ICME.

Estimates from several case studies indicated that 50 to 80 percent of the expense of developing ICME tools related to experimental investigations. As models for materials properties developed, it was often the case that earlier materials characterization was insufficient since it had been conducted primarily for the purpose of quality control. Experiments were also required to fill the gaps where theories were not sufficiently predictive or quantitative. Finally, experimental validation is critical to gaining the acceptance of the product engineering community and to ensuring that the tools are sufficiently accurate for the intended use. An important function of computational models is to capture this experimental knowledge for later reuse.

Lesson Learned 7. Databases are the key to capturing, curating, and archiving the critical information required for development of ICME.

A number of case studies, both within and outside the materials engineering discipline, highlighted the need for large-scale capture and dissemination of critical data. One showed the negative impact of failing to archive data in a recoverable form.[34] To create and utilize accurate and quantitative ICME tools, engineers must have easy access to relevant, high-quality data. Both open-access and proprietary databases permit the archiving and mining of the large, qualified, and standardized data sets that enable ICME.

Lesson Learned 8. ICME activities are enabled by open-access data and integration-friendly software.

Integration of computational models requires the transfer of data between models and customized links—that is, input and output interfaces—within models. To enable data transfer, information must be stored in accessible, standardized formats that can interface with various models. To facilitate model input and output, software must be designed to allow easy integration of user-developed subroutines, preferably through the use of open architectures that enable plug-in applications. While many of the main commercial codes used in design analysis allow user-definable subroutines, manufacturing simulation codes vary greatly in that capability and in the sophistication of the interfaces. To make fullest use of ICME, both databases and model software must be designed with open integration in mind.

Lesson Learned 9. In applying ICME, a less-than-perfect solution may be good enough.

Several case studies emphasized that ICME can provide significant value even if

---

[34]Jonathan Zimmerman, Sandia National Laboratories, "Helium bubble growth during the aging of Pd-tritides," Presentation to the committee on March 13, 2007. Available at http://www7.national academies.org/nmab/CICME_Mtg_Presentations.html. Accessed February 2008.

it is less than 100 percent accurate. While scientists may focus on perfection, many existing theories, models and tools are sufficiently well developed that they can be effectively integrated into an ICME engineering methodology. Sensitivity studies, understanding of real-world uncertainty, and experimental validation were key to gaining the acceptance of and value from ICME tools with less than 100 percent accuracy. To judge what is a reasonable balance between efficiency and robustness requires a team with expertise.

Lesson Learned 10. Development of ICME requires cross-functional teams focused on common goals or "foundational engineering problems."

All the successful ICME efforts discussed in the report were carried out by cross-functional teams made up of experts in materials, design, and manufacturing who were well versed in technology integration and had a common goal. Since many of the required tools are still under development, both engineering and research perspectives must be represented in ICME teams. Successful ICME required integration of many kinds of expertise, including in materials engineering, materials science, mechanics, mechanical engineering, physics, software development, experimentation, and numerical methods. An important part of this lesson is the selection of a common goal (such as a foundational engineering problem) that includes (1) a manufacturing process or set of processes, (2) a materials system, and (3) an application or set of applications that define the critical properties and geometries.

3

Technological Barriers: Computational, Experimental, and Integration Needs for ICME

Chapter 2 provided case studies that show the significant economic and competitive benefits that U.S. original equipment manufacturers (OEMs) and other manufacturers have achieved through the use of ICME. Those studies illustrate the integration of materials knowledge into component manufacturing, optimization, and prognosis. However, there remain significant technical barriers to the widespread adoption of ICME capabilities. In this chapter the committee discusses many of those challenges, focusing not only on modeling tools but also on the materials databases and experimental tools needed to make ICME a reality for a broad spectrum of materials applications. Finally, ways to integrate the various tools and data into a seamless ICME package are addressed.

CURRENT COMPUTATIONAL MATERIALS SCIENCE TOOLS

Today's materials scientists have increasingly powerful computational tools at their disposal. A recent DOE study demonstrates the compelling nature of the opportunities in computational materials science (CMS).[1] A recent NSF report focuses on the cyberinfrastructure needed for materials science.[2] The power of

---

[1]Department of Energy (DOE), *Opportunities for Discovery: Theory and Computation in Basic Energy Sciences* (2005). Available at http://www.sc.doe.gov/bes/reports/files/OD_rpt.pdf. Accessed February 2008.

[2]National Science Foundation (NSF), *Materials Research Cyberscience Enabled by Cyberinfrastructure* (2004). Available at http://www.nsf.gov/mps/dmr/csci.pdf. Accessed February 2008.

twenty-first century computing is making it possible to predict a range of structural features and properties from fundamental principles. These tools are diverse and range from the atomic level to the continuum level and from thermodynamic models to science-based property models. Current computational materials methods range from the specialized materials modeling methods that are used in fundamental research to the full-scale materials processing tools at manufacturing facilities. Researchers in materials science, mechanics, physics, and chemistry explore materials processing–structure–property relationships as a natural part of the research process. The results from these explorations are often incorporated into sophisticated modeling methods focused on a narrow part of overall materials behavior. While these isolated CMS methods do not necessarily contribute to the ICME infrastructure, they represent a vast supermarket of method development that can be drawn on by yet-to-be developed integration efforts and infrastructures.

The wide range of CMS methods available today are both a blessing and a curse to materials and engineering design teams. It is difficult for scientists and engineers to judge the efficacy of new or even well-established computational methods because the tools used are typically developed in somewhat isolated research environments. While this approach encourages creativity and innovation, it also means that use of these tools requires well-trained specialists who can maintain and run what are basically research codes. In other fields the computational methods—for example, finite element analysis (FEA) and finite difference methods—are firmly embodied in standard packages that have become an integral part of the academic training of the modern scientist or engineer, being based on the mathematical foundation of the discipline. In materials science and engineering, however, the scope is extremely broad and is based on a wide range of mechanisms that typically operate at different length and temporal scales, each of which needs to be modeled with specialized methods.

The properties of materials are controlled by a multitude of separate and often competing mechanisms that operate over a wide range of length and time scales, The committee concludes that since there is no single overarching approach to modeling all materials phenomena, the widespread application of materials modeling has been limited and has impeded the transformative power of ICME.

Most computational materials methods can be traced back to academic groups that developed these methods as part of the educational, scientific, and engineering process. A typical, but by no means universal, path to maturity would include several generations of research codes from one or more groups, which then are transitioned to applications in a government or industrial laboratory, then commercialized with or without government support. In the United States, federal support—through, for example, the Small Business Innovation Research (SBIR) and Small Business Technology Transfer (STTR) grants—has played a key role in commercializing processing and thermodynamic methods such as ProCast,

Deform, and Pandat. In the committee's judgment, federal support will continue to play an important role in incubating and transitioning new ICME methods.

Methods

The fundamental technical challenge of ICME is that materials response and behavior involve a multitude of physical phenomena whose accurate capture in models requires spanning many orders of magnitude in length and time. The length scales in materials response range from nanometers of atoms to the centimeters and meters of manufactured products. Similarly, time scales range from the picoseconds of atomic vibrations to the decades over which a component will be in service. Fundamentally, properties arise from the electronic distributions and bonding at the atomic scale of nanometers, but defects that exist on multiple length scales, from nanometers to centimeters, may in fact dominate properties. It should not be surprising that no single modeling approach can describe this multitude of phenomena or the breadth of scales involved. While many computational materials methods have been developed, each is focused on a specific set of issues and appropriate for a given range of lengths and times.

Consider length scales from 1 angstrom to 100 microns. At the smallest scales scientists use electronic structure methods to predict bonding, magnetic moments, and transport properties of atoms in different configurations. As the simulation cells get larger and the times scales longer, empirical interatomic potentials are used to approximate these interactions. Optimization and temporal evolution of electronic structure and atomistic methods are achieved using conjugate gradients, molecular dynamics, and Monte Carlo techniques. At still larger scales, the information content of the simulation unit decreases until it becomes more efficient to describe the material in terms of the defect that dominates at that length scale. These units might be defects in the lattice (for example, dislocations), the internal interfaces (for example, grain boundaries), or some other internal structure, and the simulations use these defects as the fundamental simulation unit in the calculation.

While true concurrent multiscale materials modeling is the goal of one segment of the materials community, for the foreseeable future most multiscale modeling will be accomplished by coordinating the input and output of stand-alone codes. This information passing approach has drawbacks associated with extracting information at each scale in an effective way. Also, all these approaches necessarily incorporate simplifying assumptions that lead to errors and uncertainties in derived quantities that are propagated throughout the multiscale integration. Experimental data play a key role here in defining parameters and information not available from simulations at all scales and in calibrating and validating modeling techniques.

Table 3-1 shows a variety of computational materials methods, some of them standard in ICME and others strictly research tools. The committee notes that the

TABLE 3-1 Mode or Method, Required Input, Expected Output, and Typical Software Used in Materials Science and Engineering

Class of Computational Materials Model/Method	Inputs	Outputs	Software Examples
Electronic structure methods (density functional theory, quantum chemistry)	Atomic number, mass, valence electrons, crystal structure and lattice spacing, Wyckoff positions, atomic arrangement	Electronic properties, elastic constants, free energy vs. structure and other parameters, activation energies, reaction pathways, defect energies and interactions	VASP, Wien2K, CASTEP, GAMES, Gaussian, a=chem., SIESTA, DACAPO
Atomistic simulations (molecular dynamics, Monte Carlo)	Interaction scheme, potentials, methodologies, benchmarks	Thermodynamics, reaction pathways, structures, point defect and dislocation mobility, grain boundary energy and mobility, precipitate dimensions	CERIU2, LAMMPS, PARADYN, DL-POLY
Dislocation dynamics	Crystal structure and lattice spacing, elastic constants, boundary conditions, mobility laws	Stress-strain behavior, hardening behavior, effect of size scale	PARANOID, ParaDis, Dis-dynamics, Micro-Megas
Thermodynamic methods (CALPHAD)	Free-energy data from electronic structure, calorimetry data, free-energy functions fit to materials databases	Phase predominance diagrams, phase fractions, multicomponent phase diagram, free energies	Pandat, ThermoCalc, Fact Sage
Microstructural evolution methods (phase-field, front-tracking methods, Potts models)	Free-energy and kinetic databases (atom mobilities), interface and grain boundary energies, (anisotropic) interface mobilities, elastic constants	Solidification and dendritic structure, microstructure during processing, deployment, and evolution in service	OpenPF, MICRESS, DICTRA, 3DGG, Rex3D
Micromechanical and mesoscale property models (solid mechanics and FEA)	Microstructural characteristics, properties of phases and constituents	Properties of materials—for example, modulus, strength, toughness, strain tolerance, thermal/electrical conductivity, permeability; possibly creep and fatigue behavior	OOF, Voronoi Cell, JMatPro, FRANC-3D, ZenCrack, DARWIN
Microstructural imaging software	Images from optical microscopy, electron microscopes, X-rays, etc.	Image quantification and digital representations	Mimics, IDL, 3D Doctor, Amira
Mesoscale structure models (processing models)	Processing thermal and strain history	Microstructural characteristics (for example, grain size, texture, precipitate dimensions)	PrecipiCalc, JMat Pro

Class of Computational Materials Model/Method	Inputs	Outputs	Software Examples
Part-level FEA, finite difference, and other continuum models	Part geometry, manufacturing processing parameters, component loads, materials properties	Distribution of temperatures, stresses and deformation, electrical currents, magnetic and optical behavior, etc.	ProCast, MagmaSoft, CAPCAST, DEFORM, LS-Dyna, Abaqus
Code and systems integration	Format of input and output of modules and the logical structure of integration, initial input	Parameters for optimized design, sensitivity to variations in inputs or individual modules	iSIGHT/FIPER, QMD, Phoenix
Statistical tools (neural nets, principal component analysis)	Composition, process conditions, properties	Correlations between inputs and outputs; mechanistic insights	SPLUS, MiniTab, SYSTAT, FIPER, PatternMaster, MATLAB, SAS/STAT

table is not intended to be complete but rather to exemplify the methods available for modeling materials characteristics. This table indicates typical inputs and outputs of the software and examples of widely used or recognized codes. Electronic structure methods employ different approximate solutions to the quantum mechanics of atoms and electrons to explore the effects of bonding, chemistry, local structure, and dynamics on the mechanisms that affect material properties. Typically, tens to hundreds of atoms are included in such a calculation and the timescales are on the order of nanoseconds. In atomistic simulations, arrangements and trajectories of atoms and molecules are calculated. Generally based on models to describe the interactions among atoms, simulations are now routinely carried out with millions of atoms. Length scales and timescales are in the nanometer and nanosecond regime, and longer length scales and timescales are possible in the case of molecular system coarse graining from "all-atom" to "united atom" models (that is, interacting clusters of atoms). Dislocation dynamics methods are used to study the evolution of dislocations (curvilinear defects in the lattice) during plastic deformation. The total number of dislocations is typically less than a million, and strain rates are large compared to those measured in standard laboratory tests. Thermodynamic methods range from first-principle predictions of phase diagrams to complex database integration methods using existing tabulated data to produce phase diagrams and kinetics data. These methods are being developed by the CALculation of PHAse Diagram (CALPHAD) community (see Box 3-1). Microstructural evolution methods predict microstructure stability and evolution based on free-energy functions, elastic parameters, and kinetic databases. Recently, several groups established protocols to automatically extract thermodynamic and kinetic information from CALPHAD methods as input to such methods. Micro-

BOX 3-1
CALculation of PHAse Diagrams (CALPHAD) Methodology

The calculation of phase diagrams is a well-developed and widely accepted computational approach for capturing and using materials thermodynamic information. Personal-computer-based commercial software, coupled with commercial and open databases of thermodynamic information (data and models), provides the results of sophisticated and accurate calculations. Now readily available to those with even modest backgrounds in thermodynamics and phase equilibria calculations, these thermodynamic simulations based on critically evaluated data are basic tools in materials and process design.[1] However the development of the sophisticated tools and databases in use today took more than 50 years and was the result of the efforts of countless contributors. Because CALPHAD software is arguably the most important (and perhaps the only) generic tool available for ICME practitioners, a brief examination of its history could reveal how ICME is likely to develop.[2]

Although this method is ultimately rooted in work begun in the early 1900s, the modern CALPHAD movement began in the late 1950s, when the global scientific community began to envision a phase diagram calculation capability based on extensive databases of thermodyamic properties and empirical data. Over the course of 50 years this vision was a constant goal. While the time required to bring the effort to fruition may seem long, Saunders and Miodownik have suggested that such a lengthy incubation period between vision and fruition reflects the time required for individuals to meet each other and agree to work together and the time for science and technology to dedicate adequate funds. They also suggested that a contributing factor was the difficulty some scientists had in accepting that realizing this vision required a melding of empirical databases and fundamental thermodynamics.

Since the late 1950s, many factors enabled CALPHAD to develop:

- Visionary leaders who understood the potential of CALPHAD and who worked continuously for decades to make it a reality.
- CALPHAD research groups at universities and government laboratories such as the National Bureau of Standards (now NIST), often led by the aforementioned individuals, who provided continuity and sustained effort and focus.
- A strong community of experts.
- Technical conferences dedicated to CALPHAD that enabled researchers to interact and collaborate.
- A focus on practical problems of interest to industry—for example, steels and nickel-based superalloys.
- Textbooks dedicated to CALPHAD (the first was published in the 1970s).
- Establishment of a journal (in 1977) dedicated to publication of CALPHAD data.
- International agreements and international consortia dedicated to the CALPHAD vision—one such is the Scientific Group Thermodata Europe (SGTE)—have been in existence since the 1970s.
- Substantial public funding of database development especially in Europe via the program Cooperation in the Field of Scientific and Technical Research (COST).

---

[1]P.J. Spencer, ed., "Computer simulations from thermodynamic data: Materials production and development," *MRS Bulletin* 24(4) (1999).

[2]For a more comprehensive review of the history of CALPHAD, see N. Saunders and A.P. Miodownik, *CALPHAD—Calculation of Phase Diagrams, A Comprehensive Guide*, Oxford, England: Elsevier (1998).

- Expert practitioners who develop phase diagram assessments and make them available to others either freely or via commercial databases linked to commercial CALPHAD software.
- The use of common thermodynamic reference states along with a shared and agreed-on taxonomy.
- The open publication and sharing of common data (at least for unaries and often for binaries and ternaries) that form the building blocks for many of the CALPHAD databases.
- PC-based commercial software and databases that can be operated without extensive expertise.
- Commercial software with programming interfaces that enable users to write their own applications and call up key functions on demand.

Current CALPHAD development efforts include establishment of linkages with physics-based tools such as density functional theory for calculating the energetics required to assess phase stability and linkage with and development of diffusion databases and models that are in turn linked to microstructural evolution prediction tools. Finally, some developers of CALPHAD tools have begun to venture into property prediction, by either correlations or science-based models, setting the stage for the use of CALPHAD as a basic ICME tool. The enabling factors that led to the CALPHAD capability of today will also be critical enablers for the development of a widespread ICME capability in the future.

mechanical and mesoscale property models include solid mechanics and FEA methods that use experimentally derived models of materials behavior to explore microstructural influences on properties. The models may incorporate details of the microstructure (resolving scales at the relevant level). Results may be at full system scale. *Mesoscale structure models* include models for solidification and solid state deformation using combinations of the previous methods to predict favorable processing conditions for specific microstructural characteristics. Methods for code and systems integration offer ways to connect many types of models and simulations and to apply systems engineering strategies. Statistical tools are often used to gain new understanding through correlations in large data sets. Other important ICME tools include databases, quantifiable knowledge rules, error propagation models, and cost and performance models. To be effective in an ICME environment, all of these computational methods must be integrated with other tools. Developing such compatibilities should be a priority for model developers and funding agencies. The development of standards and common nomenclatures for data exchange and model compatibility is an important task and is discussed in more detail in the sections "Requirements for ICME Databases" and "Commercial Integration Tools."

It would be beyond the scope of this report to give details of the advances that are needed for all the methods employed to model materials behavior. Table 3-1 lists methods along with their inputs and outputs. While each method is by itself a critical component of an ICME process, linking the various methods remains a great challenge, not only from a scientific perspective but also because the codes for these models may exist on different computer platforms and be written in different languages. While still an unsolved problem, projects like those sponsored by Eclipse are focused on the creation of open development platforms to make such computational linkages easier.[3] Each class of methods in Table 3.1 has its own needs and challenges, among them the following:

- Extensions of atomistic simulations to longer times through the use, for example, of accelerated dynamics methods and to broader classes of materials systems through the development and validation of force fields for application to heterogeneous/mixed materials, especially at the interfaces between material types (for example, metal-ceramic);
- Development of spatially hierarchical microstructural evolution methods for concurrently modeling microstructural features across length scales;
- Advances in crystal plasticity finite element methods to include the effects of local heterogeneities in the microstructure;
- Methods for modeling the spatial and temporal scales between dislocation dynamics and continuum level (for instance, finite element methods);
- Science-based models for predicting the influence of microstructure on a wide variety of properties.
- Development of improved microstructural evolution models for polymers, polymeric composites, and elastomers;
- Advances in electronic structure calculations for modeling larger systems (for example, development of spatially hierarchical methods employing a flexible—such as a wavelet—basis) and for more accurately accounting for electron correlation, which will be critically important for materials at the nanoscale; and
- Development of diffusion data and kinetics theory to explain a wide variety of materials phenomena in metals, polymers, and ceramics.

This list, while far from complete, indicates the diversity of challenges in computational materials. For ICME, the key is to influence the directions these developments take, with the goal being greater integration between models for different materials phenomena and across scales and better integration of data within the models and simulations.

---

[3]For more information, see http://www.eclipse.org. Accessed February 2008.

Advances in Computing Capabilities

ICME is possible today in part because of the exponential growth in computer storage and processing capability achieved over the last 40 years. Current desktop processors yield performance reserved for the supercomputers of a decade ago, multigigabytes of memory have become standard, and disks can store terabytes of information, all at an affordable cost. Thus the computational capabilities required to model materials behavior for ICME are becoming increasingly available to the practicing materials engineer. The prognosis is for continued improvements in hardware capabilities.

The recent advances in multiprocessor computing have had a dramatic effect on the utility of a variety of methods, with a natural evolution of computational methods from serial to scalable parallel processing. The stages to full parallel processing include using parallel processing compilers, "parallelizing" computation-intensive portions of the code, "parallelizing" the original serial implementation, and redesigning the serial implementation to take full advantage of the available parallel architectures. Many commercial applications (for example, finite element methods) are available for parallel computing and are in common use in industrial settings. With the focus on multicore processors from the computer vendors, computing methods that take advantage of parallelism and that will require new and different programming paradigms will become increasingly common.

Tools for ICME will need to have a number of features to take full advantage of the power of modern and evolving computing platforms. While the technical details are beyond the scope of this report, successful methods will generally include the following:

- *Scalable parallelism.* As the cost of processors continues to fall, the ability to scale to hundreds or thousands of processors will be paramount.
- *I/O and file systems.* Many classes of simulation tools are constrained by communication bandwidth between processors and to storage servers. For example, just reading the results of the Los Alamos National Laboratory simulations shown in Figure 3-1 required 100 servers. Higher computing performance will require new algorithms that scale without such heavy I/O burdens. New distributed file systems can help significantly with the storage and retrieval of large amounts of data, and materials simulations and visualizations will need to work well with these file systems.
- *Advances in graphics hardware.* New graphics processing units (GPUs) can offer very large sustained processing speeds (up to 50 Gflop as this report is written), which is considerably faster than general-purpose central processing units. The general material application development community has done little to take advantage of this technology; however, graphics-

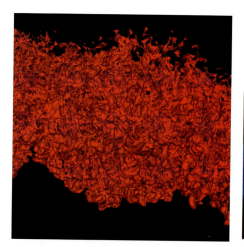
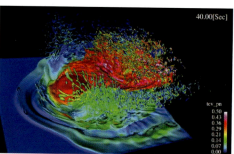

FIGURE 3-1 *Left:* Shock turbulence model with 589 million elements rendered at an effective rate of 3.2 billion polygons per second on 128-pipe Army Research Laboratory visualization system. SOURCE: Lawrence Livermore National Laboratory. *Right:* Asteroid impact study—240 million cells, 9.7 TB, 50 servers, multiple angles. Image courtesy of CEI.

hardware-based computing (gpgpu) has been widely used in some massively parallel astrophysics applications. The next generation of gpgpu hardware appears to hold great promise for high-performance computing. New materials and applications in biology might benefit from this technology.

- *Intelligent data reduction.* It is relatively easy to use 10,000 processors for a large, highly scalable computational fluid dynamics (CFD) application or to start thousands of design variations. It is more difficult to ensure the timely delivery of input data or the creation of large output files for large simulations that are run in parallel. Using such large amounts of data will necessitate the intelligent reduction of information required for the next-higher level of integration of materials or systems models.

- *Fault detection and recovery.* In a cluster with thousands of processors, the mean time to failure of a single processor is less than 1 day. Simulation tools will thus need robust fault detection and recovery capabilities. Low-level fault detection is just entering compilers and parallel computing middleware such as message-passing interface (MPI), but no broadly available materials simulation tools currently take advantage of these capabilities or provide their own fault detection to improve their reliability.

- *Out-of-order execution.* Modern processors with large numbers of cores and threads per core can run much faster when programmed such that most instructions can be scheduled either simultaneously or out of order. No broadly available materials simulation tools are written to take advantage of such capabilities.

- *Petaflop computing.* Next-generation computers, capable of petaflop performance, will probably employ hundreds of thousands of processors. New programming paradigms will be needed to achieve scalability on these massive machines.

Computational capabilities will continue to increase, enabling higher fidelity and complexity in ICME applications. By taking advantage of new architectures and software enhancements, application developers can enhance the ability of their computational tools to meet the challenges listed in the preceding section. The committee notes, however, that few scientists and engineers in the materials community have the training to fully engage in the development of modern computational methods, so collaboration with computer scientists will be essential. Accordingly, institutions that engage in materials education, development, and manufacturing will need to undergo significant cultural change; this is discussed in Chapter 4.

Uncertainty Quantification

ICME requires the development of predictive models and simulations, the quality (accuracy and precision) of whose final results will have to be known by the materials engineer.[4] The ability of the ICME community to predict the quality of a coupled set of calculations is limited, because there can be considerable uncertainty at almost all levels of the ICME process. All materials models have uncertainties associated with the natural variability of materials properties that arises from the stochastic nature of materials structures. The problem is exacerbated by the critical dependency of many materials properties on the distribution of defects (that is, on microstructural heterogeneities), which are, in turn, influenced by processing variables. Thus it is very important to carefully calibrate and validate modeling tools by comparing their results to the results of well-designed experiments on pedigreed materials. Beyond the uncertainties in the materials models, all simulation methods have their own levels of uncertainty, from the stochastic uncertainty of a molecular dynamics simulation to the numerical uncertainty of a large-scale finite element calculation. A key need for all ICME applications is quantification of uncertainties in each stage of a suite of calculations.

---

[4]Since ICME is best practiced with complementary experimental and theoretical approaches, the validation of computational methods to fill gaps in theoretical understanding is critical to building a robust ICME approach. Validation is discussed in the section "Role of Experimentation in Computational Materials Science and ICME."

The quantification of uncertainty is a burgeoning field within computational science and engineering. Many methods have been developed to track how uncertainties in input parameters and the underlying models propagate through a simulation—that is, how input affects output. There are two general approaches, sensitivity analysis and uncertainty quantification. In sensitivity analysis, one examines how changes in inputs affect outputs. Uncertainty quantification is a more global process, which involves a statistical analysis based on probability distributions. In both approaches, input parameters are treated as stochastic variables, with an associated probability distribution function. Various statistical methods are then employed to estimate the distribution of outputs associated with those inputs. Full integration of uncertainty quantification into ICME processes will be essential for developing the reliable applications needed for widespread acceptance of ICME in the integrated product development (IPD) process and for more efficient development of future ICME tools.

Visualization

The ability to visualize the output of complex analysis is crucial to understanding design challenges of today and tomorrow. Advances in commodity graphics hardware have led to the replacement of virtually all of the graphics hardware of the once-dominant proprietary graphics vendors. Yet even the most powerful graphics cards are not able to process large-scale models needed to accurately perform complex analysis. Moderately sized models can be loaded into a single card but cannot be rendered fast enough to provide meaningful interaction with the data. In the CFD codes that simulate manufacturing processes such as the casting of metals or the injection molding of plastic, the number of cells or elements exceeds 10-15 million, and additional graphics capability is required. Software development efforts have now created parallel-distributed graphics applications codes that can take advantage of graphics hardware installed on individual nodes of a compute cluster. Examples of these codes are EnSightDR, Paraview, and Chromium. These distributed nodes, combined with a fast interconnect and high-performance I/O, give users the ability to interact with large models. Rather than being limited to images generated from, say, the x, y, and z axes only, the interactivity of these distributed systems allows users to zoom, pan, rotate, and display time-dependent animations on models ranging from 50 million to 1 billion elements or cells.

With advanced visualization tools, the time-dependent results of large models showing changing states and structures can be more easily understood. Figure 3-1 illustrates a large shock turbulence model, which was rendered on a large Army Research Laboratory visualization cluster. The model is composed of 589 million polygons. Using an off-the-shelf production version of EnSightDR, the model can be updated by six frames per second. This gives researchers the ability to interactively examine localized phenomenon anywhere on the geometry.

One of the difficulties with any collaborative effort is obtaining access to compute and visualization resources. Boeing recently conducted a series of benchmark tests to validate the usefulness of remote visualization of moderate-size models. To test the performance of remote graphics, a 60 million cell model was rendered in Bellevue and displayed in Renton. The resulting images were updated at 10 frames per second, which is more than adequate for engineering design review and post processing. The conclusion was that remote visualization of moderately large models is now feasible, which saves large data transfers and duplicate disk space on both the computing platform and the graphics platform.

One last challenge in visualization is to represent these very large data sets in ways that are comprehensible to the human mind. A computer may be able to display the results of a simulation with a billion degrees of freedom, but our ability to understand and use the data will depend on the visualization designer's ability to highlight the important features in the result. This is true of high-performance physical simulations in general: visualization challenges specific to materials include electron orbital interactions in large density functional calculations; large ensembles of dislocations; crazing, crystallinity, and other complex structures in polymers; and distributions of structure features such as precipitates, dendrites, and grains throughout an engineering component.

ROLE OF EXPERIMENTATION IN COMPUTATIONAL MATERIALS SCIENCE AND ICME

Experimental Calibration and Validation

One of the important lessons of earlier ICME efforts is the profound importance of experimental results for calibrating and validating computational methods and filling gaps in theoretical understanding. In fact ICME is best practiced with complementary experimental and theoretical approaches. This implies well-integrated research teams working toward a common goal, such as would be found in industry and government laboratories. Based on the lessons learned from the DOE advanced simulation and computing initiative (ASCI), verification and validation of modeling methods has become a key part of the current National Nuclear Security Administration (NNSA) Predictive Science Academic Alliance Program. Using the correct mathematical description and numerical representation (verification) and ensuring that the results are consistent with well-designed experiments (validation) should also become standard practice in the ICME community.

There are three classes of experimental effort:

- *Classical experimental validation, such as thermodynamic measurements and experiments designed to advance mechanistic understanding.* These disciplines are often poorly funded, because such mature methods are incorrectly perceived as unimportant research areas.
- *Novel experimental techniques, such as three-dimensional materials imaging microscopy and miniature sample technology.* These are cutting-edge research areas where the methods have not yet matured or permeated the materials science and engineering community.
- *High-throughput techniques, such as combinatorial materials science.* These are new and unproven but have the potential to rapidly populate databases and enable large-scale ICME.

For an ICME strategy to be successful, a strong link between experimental data and modeling is essential. Paradoxically, significant advances in low-cost processing–structure and structure–property experimental methods may be required to advance the modeling and simulation infrastructure.

Three-Dimensional Microstructural Characterization

Structure–property models for materials are at the core of any ICME implementation. Virtually all engineering materials contain structural features at the micro- or nanoscale that strongly influence properties. These structural features are in turn strongly influenced by the manufacturing processes used for a particular product and may include grain boundaries, second phases, pores, or defects such as dislocations and vacancies. They are often irregularly shaped and distributed nonuniformly in three dimensions. Knowledge of these features is particularly important for predicting flaw-sensitive properties such as fatigue. Most traditional characterization techniques collect information from two-dimensional sections and do not accurately capture the structural complexity. Thus three-dimensional images are often needed to extract quantitative structural information for property models. An example of a precipitate with a highly complex dendritic shape in three dimensions is shown in Figure 3-2.

Three-dimensional imaging is of importance in a number of other technical fields, with medical imaging being a notable example. Imaging modalities such as X-ray computed tomography, magnetic resonance imaging, ultrasound, and positron emission tomography are now routinely used as diagnostic tools. These tomographic techniques acquire sequential two-dimensional "slices" of the object of interest and digitally reconstruct these slices into a three-dimensional data set

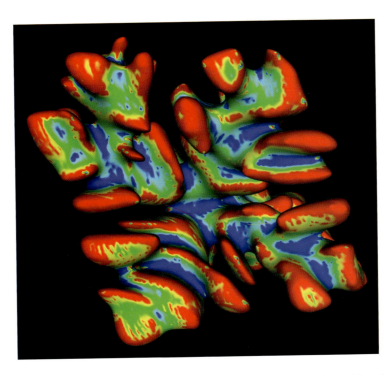

FIGURE 3-2 Three-dimensional rendering of a dendritic precipitate obtained by serial sectioning and reconstruction. Image courtesy of M. De Graef, Carnegie Mellon University.

that can be analyzed in detail. There are hardware, software, and data challenges in three-dimensional imaging. Imaging sources and detectors, automated stages, and focusing or slicing techniques must be matched to the material being examined (soft tissue, bone, metallic, polymeric, ceramic). Data sets can be quite large, challenging the processing power of computers used for meshing, reconstruction, and analysis. Image-processing routines unique to the tomography technique or class of material are often needed.

Unlike such imaging in the medical field, the three-dimensional imaging of engineering materials is not yet widely available or well developed. By their nature, engineering materials are not amenable to many of the imaging modes utilized in the medical field. Except for special cases like synchrotron radiation, where large volumes of material are transparent to the imaging radiation, engineering materials must often be physically sectioned to acquire two-dimensional imaging "slices." Recent interesting serial sectioning approaches include automated robotic serial sectioning,[5] focused ion beam sectioning,[6,7,8,9,10] and the three-dimensional atom probe.[11] Improvements in the efficiency of these techniques as well as completely new techniques are needed to bring the materials community to the point where three-dimensional information can be routinely acquired. While continued advances in computing capabilities will mitigate some of the difficulties of processing large data sets, there is no agreement on three-dimensional data formatting standards nor are any community sites available for storage of these large data sets. Utilization of three-dimensional data in property models requires new quantitative analytical approaches beyond simple two-point correlations. Under development are automated protocols for representation and stereological analysis, meshing of the complex geometrical details of the microstructure, and direct linkage to finite element analysis for bridging to macroscopic properties.[12] Given the magnitude of the three-dimensional imaging and analysis task for the materials community, significant coordination will be needed.

Rapid, Targeted Experimentation

As described above, the availability of experimental data to fill gaps in theoretical understanding, calibrate models, and validate ICME results is a key prerequisite for widespread ICME utilization. While experimental evaluation of new materials and evaluation of the influence of new processing approaches have been the cornerstone of materials development, such evaluation is slow and often very expensive, so that a more rapid approach to experimental exploration is needed. While R&D literature contains a great deal of useful information that could be harvested, newly

---

[5]J.E. Spowart, H.H. Mullens, and B.T. Puchala, "Collecting and analyzing microstructures in three dimensions: A fully automated approach," *Journal of The Minerals, Metals & Materials Society (JOM)* 36 (October 2003).

[6]B.J. Inkson, M. Mulvihill, and G. Mobus, "3D determination of grain shape in a FeAl-based nanocomposite by 3D FIB tomography," Scripta Materialia 45(7): 753-758 (2001).

[7]M. Groeber, B. Haley, M. Uchic, and S. Ghosh, Materials Processing and Design, *NUMIFORM 2004*, American Institute of Physics Conference Proceedings 712.

[8]A.J. Kubis, G.J. Shiflet, D.N. Dunn, and R. Hull, "Focused ion-beam tomography," *Metallurgical and Materials Transactions A* 35(7): 1935-1943.

[9]Y. Bhandari, S. Sarkar, M. Groeber, M.D. Uchic, D.M. Dimiduk, and S. Ghosh, "3D polycrystalline microstructure reconstruction from FIB generated serial sections for FE analysis," *Computational Materials Science* 41(2): 222-235.

[10]A. Shan and A.M. Gokhale, "Digital image analysis and microstructure modeling tools for microstructure sensitive design of materials," *International Journal of Plasticity* 20(7): 1347-1370.

[11]S.S.A. Gerstl, D.N. Seidman, A.A. Gribb, and T.F. Kelly, "LEAP microscopes look at TiAl alloys," *Advanced Materials and Processes* 10 (October): 31-33.

[12]G. Spanos, ed., "3-D characterization and analysis of materials," *Scripta Materialia Viewpoint Set* 55(1) (2006).

developed materials or new processing routes inevitably call for experimental data for the calibration and validation of models. There are emerging suites of new characterization tools that permit materials properties to be rapidly screened and evaluated without the need for large volumes of material. An example of this is the diffusion multiple approach. With a single sample containing pieces of various pure elements bonded together (Figure 3-3), aspects of binary and ternary phase diagrams can be explored with the use of a single sample, reducing by a factor of 20 to 100 the required number of samples that must be processed to obtain this fundamental information. Among the new techniques for probling properties in small volumes are local laser-based probes for thermal and electrical conductivity, nanoindentors, and new electron backscattered scanning electron microscopy (SEM) techniques. The efficiency of such approaches can greatly accelerate the materials development process while reducing its often extraordinarily high cost. Figure 3-4 illustrates the use of focused ion beam milling with microcompression methods to sample the local micron-scale properties.[13,14] Such techniques can be used to sample the effects of microstructural heterogeneities across a component or to screen the properties of small volumes of new materials without having to develop or utilize high-volume, time-consuming, and expensive processing operations. Currently these methods are limited by the milling time required to produce the samples, but alternative methods for producing massive arrays of microsamples are under development. Other thin- and thick-film combinatorial processing approaches, rapid micromachining property evaluation, and microscale mechanical tests are among the suite of emerging tools that promise to dramatically accelerate the materials and product development cycles. These techniques are still in their infancy, and protocols for acquiring, storing, and sharing the vast amounts of data they might generate have yet to be developed.

DATABASES AND ICME DEVELOPMENT

Over the course of the study, databases emerged as important enablers in the ICME infrastructure. Databases provide a mechanism for storing experimental and computational results and for efficiently linking to models operating at different length scales or timescales. As shown in Figure 3-5, the diversity of data types required for complete knowledge of a material necessitates a variety of database structures to meet the input needs of the various models. Currently the schemas for describing the breadth of database types are at varying degrees of maturity. This

---

[13]M.D. Uchic, D.M. Dimiduk, J.N. Florando, and W.D. Nix, "Sample dimensions influence strength and crystal plasticity," *Science* 305: 986-989 (2004).

[14]Q Feng, Y.N. Picard, H. Liu, S.M. Yalisove, G. Mourou, and T.M. Pollock, "Femtosecond laser micromachining of a single-crystal superalloy," *Scripta Materialia* 53(5): 511-516 (2005).

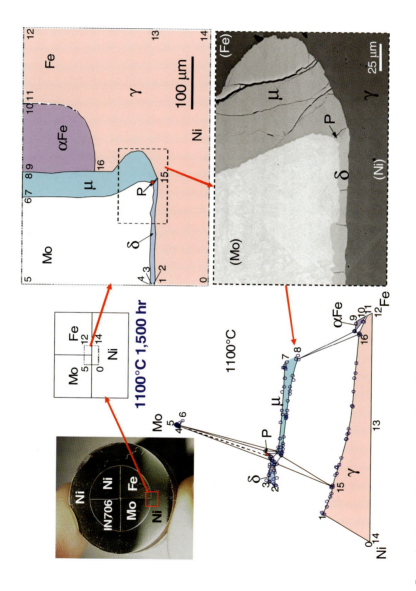

FIGURE 3-3 Example of a combinatorial approach: application of a diffusion multiple to generate data on single-phase compositions of a ternary system for a survey of Ni-Mo-Fe. A diffusion multiple made up of Ni, Fe, Mo, and superalloy Inconel 706 (IN706) for rapid mapping of the Ni-Fe-Mo phase diagram and for studying the alloying effect of Mo addition to IN706. The phase diagram is plotted using atomic percent axes with numbering of the scales removed for simplicity. SOURCE: J.C. Zhao, "The diffusion-multiple approach to designing alloys," *Annual Review of Materials Research* 35: 51-73 (2005).

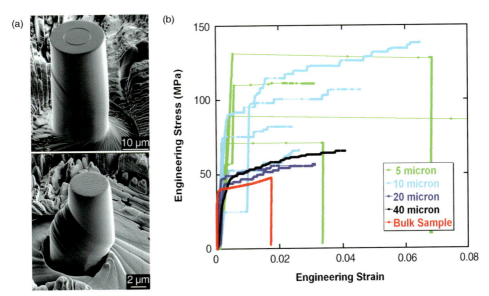

FIGURE 3-4 Locally machined micropillars (a) fabricated in a focused ion beam system. They are compressed with a nanoindentor tip to obtain stress-strain data (b). SOURCE: M.D. Uchic, D.M. Dimiduk, J.N. Florando, and W.D. Nix, "Sample dimensions influence strength and crystal plasticity," *Science* 305 (5686):986-989 (2004). Reprinted with permission from AAAS.

section reviews the current state of databases useful to ICME, explores those data requirements, and anticipates the ways in which databases will need to expand in order to accommodate the needs of ICME models.

Current Status and Database Issues

When databases are properly constructed and maintained, they empower the efficient use of materials data in systems design by IPD processes. Progress in establishing such databases in materials science and engineering is hindered by several problems. Often it is difficult to delimit the scope of the data to be stored or to determine their nature, because the mechanisms controlling a given property of a material may not be known. This also requires that the data, which can take many forms (numbers, images, graphs), be stored in a compact but low-loss procedure so that they can be resampled in the future. The variety of communities accessing and contributing to these databases means the databases must be both transparent and secure. Finally, while the manufacturing industry has the most to gain from ICME, the companies have competitive reasons for not moving significant parts of their materials knowledge base into the public domain. Many aspects of databases (for

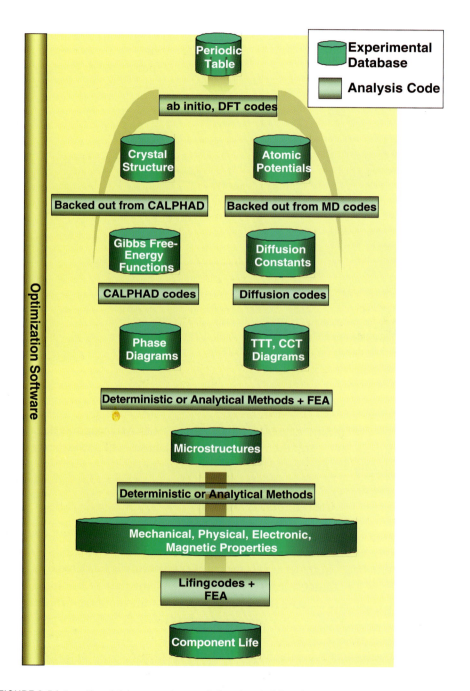

FIGURE 3-5 Integrating databases and computational materials science tools. DFT, density functional theory; MD, molecular dynamics; TTT, time–temperature transformation curve; and CCT, continuous cooling transformation curves.

example, funding, formatting, administration, populating, and business support) will need to evolve in order to better serve the emerging ICME infrastructure.

In several areas of ICME small companies license powerful but low-cost drivers and graphical user interfaces (GUIs) that allow access to large proprietary databases. Such is the case for the materials thermodynamics and phase diagrams (CALPHAD: CALculation of PHAse Diagram) community, which constructs databases from literature data and unpublished academic and industrial studies and then sells them as part of a thermodynamic modeling package. Other companies—such as MatWeb and Granta—offer large databases of material properties in the form of a commercial product or as inexpensive Web-based products supported by selling advertising space on the Web site, much like the commercial search engines Yahoo and Google. Because properties are so dependent on processing and structure, property databases that contain little or no information on manufacturing history or microstructure are, in the committee's judgment, of limited utility for development of ICME. Since ICME seeks to build a tool set that links material composition and structure with material properties, such databases must contain lower-level inputs (for example, crystal structures, thermodynamics data, kinetic data, and physical properties). This is analogous to the bioinformatics problem, where a record on a nucleotide DNA subunit within a database would, for instance, contain information on the organism from which it was isolated, the input sequence, type of molecule, and a literature citation.[15]

Some materials professional societies play an important role in database development by bringing together industry, government, and academia to address data classification issues and by hosting materials properties databases. For example, ASM International has partnered with Granta, a privately held company based in the United Kingdom, to launch a Web-based materials information center. The site offers databases and software products for a variety of industries ranging from medical devices to aerospace and defense. Materials data can be managed through the group's proprietary software (GRANTA-MI), and another GUI enables materials and process selection using the Cambridge Materials Selector. This software incorporates a variety of standard databases relevant to different communities—for example, MMPDS data (previously MIL-HDBK-5), which is maintained by the Federal Aviation Administration for aerospace applications. Unfortunately these precompetitive databases—data in the public domain—are quite limited compared with what an IPD team in an industrial manufacturing setting actually needs.

Investment in materials databases is at a very low level compared to investment in biological and genomics databases. Significant government support of genomic databases coincided with the realization that such data were sufficiently important to growth in this area that any collected data should immediately become part of

---

[15]For more information, see http://www.ncbi.nlm.nih.gov. Accessed February 2007.

the community's common knowledge base through archival, curated Internet databases. This approach has been enforced by requiring the original authors to post their data before publishing their findings. Such a requirement forces researchers to meet a certain level of fidelity in their data gathering and enforces a common data structure across the community. Currently GenBank, housed at the National Center for Biotechnology Information (NCBI),[16] has over 130 gigabases of sequences, and Entrez[17] offers an integrated database of databases allowing text-based searches. Smaller, model organism databases that house genome sequences and data on structure and function (such as FlyBase,[18] WormBase,[19] Mouse Genome[20]) cost between $400,000 and $1 million per year to develop and curate and are generally funded by the government through, for instance, the National Institutes of Health (NIH). Significant government investments, similar to those made by the NIH in the genomics community, will be required to create and curate the precompetitive databases required to support ICME.

Requirements for ICME Databases

Those who have built large scientific databases in the biology and physics communities emphasize two design principles.[21,22] First, useful databases must start with a taxonomy of the field that is comprehensive and able to accommodate change. The fundamental problem with materials databases is that their structure, or schema, is generally focused narrowly on the immediate problem of the user base and is not easily expandable. Second, one must anticipate as much as possible what questions people will want to ask, because today's data representation may not be flexible enough to suit future lines of inquiry. The development of materials and their transitioning to applications require efficient, informed, and flexible descriptions of the salient measurables driving property variability across a component.

For example, one key measurable that presents particular challenges in stor-

---

[16]For more information, see http://www.ncbi.nlm.nih.gov/. Accessed December 2007.

[17]For more information, see http://www.ncbi.nlm.nih.gov/sites/gquery. Accessed December 2007.

[18]For more information, see http://flybase.bio.indiana.edu/. Accessed December 2007.

[19]For more information, see http://www.wormbase.org/. Accessed December 2007.

[20]For more information, see http://www.ncbi.nlm.nih.gov/genome/guide/mouse/. Accessed December 2007.

[21]Cate L. Brinson, Northwestern University, "Materials informatics–what, how and why: Analogy to bioinformatics," Presentation to the committee on May 30, 2007. Available at http://www7.national academies.org/nmab/CICME_Mtg_Presentations.html. Accessed February 2008.

[22]Krishna Rajan, Iowa State University, "Materials informatics," Presentation to the committee on March 13, 2007. Available at http://www7.nationalacademies.org/nmab/CICME_Mtg_Presentations. html. Accessed February 2008.

age and retrieval is the nature and variation in microstructure. There are ongoing materials science efforts to build microstructural databases, particularly for structural metals. The Office of Naval Research program on dynamic digital three-dimensional structure is exploring methods for collecting, storing, and retrieving microstructural data in the three-dimensional Materials Atlas. Microstructural data, images, and metadata are compressed using low-loss techniques and stored using a relational database (SQL) and a data format (HDF5) that were chosen for robustness and flexibility. The long-term success of the Materials Atlas will be determined by the ease with which users can select, retrieve, decompress, and use the information in as-yet-to-be-determined applications and the availability of resources for curation.

The breadth of the databases required for ICME and the need for efficient data acquisition suggest the adoption of standardized formats. Such data and their repositories should facilitate the prediction of macroscopic materials properties (including cost) to meet the needs of design, manufacture, and prediction of lifetime and reliability. They must allow the efficient passing of data between models at different length scales, from different domains, and at times between different institutions (see Figure 3-5 and other examples in this section). Database designers are likely to use different approaches to pass and store information, and there are many such formats in use today, most of them geared to a particular modeling tool. Linking databases to models requires data communication via standard formats.

In order to consolidate the collection of formats for exchanging data on the properties of materials, the National Institute of Standards and Technology (NIST) created the XML interchange format called MatML. The format was designed to be very flexible in specifying properties, limitations, uncertainties, sources, and generation methods. Unfortunately, failure to specify nomenclature makes MatML too flexible for machine reading, and there is no clear way to write a universal MatML importer for use in a mechanics code. For example, there are several possible ways to express the name of the elastic modulus and different meanings for the term "modulus" in mechanics. To manage these problems, Japan's National Institute for Materials Science created a taxonomy and then defined a subset of property names to be used within MatML. Unfortunately the complexity of implementing this standard remains a barrier to its widespread adoption.

Advances in ICME will be strongly dependent on the balance struck between transparency and security in precompetitive databases. To produce relevant and reliable simulation and experimental tools, the academic community and the government laboratories require some level of transparency. However, to maintain their competitive advantage, OEMs may decide to maintain proprietary databases for their core technologies. Ideally, ICME software should be able to seamlessly integrate multiple proprietary and public data sources as inputs to materials and system integration tools in order to optimize over all the available data. Significant growth in the precompetitive databases and a well-thought-out security strategy will be required to create and maintain this part of the ICME infrastructure. Models, untied from their proprietary databases, should be released into the public domain.

One of the lessons learned from previous ICME efforts is the profound importance of having experimental results to fill gaps in theoretical understanding, so that a strong link between experimental data and modeling is essential for an ICME strategy to be successful. For those data to be accessible, the community needs a set of common databases, in much the same way that the genetics community requires a database of sequences of genes. It is not enough, however, to have access to data. Materials development requires an understanding of how different features in the data may be correlated with others. For making those connections, the committee turns to a new field, materials informatics, based on ideas from the biological community.

Materials Informatics

A major challenge facing efficient development and widespread adoption of ICME is to develop a better linkage between experimental data and modeling. Materials informatics, a promising new development in materials research, offers a way to meet that challenge. Materials informatics employs creating databases and advanced data mining and analysis methods to seek patterns of behavior from large, complex data sets, with the goal of identifying new physical relationships between chemistry, structure, and properties. Because the data sets used in informatics are not restricted to experimental data, materials informatics has potential for providing a means to connect the results of calculations and models with experiment to decrease the time required to develop a new material model or a new material.[23]

The advantage of informatics is that it makes it possible to extract information from large and complex data sets. A number of methods have been developed to mine such information. One such method, principal component analysis, allows a

---

[23]Because materials knowledge is embodied in experimental data, physically based models, empirical rules, and heuristics, the fusion of information from all available sources will increase the level of confidence in ICME analysis results. Although methods to fuse data and modeling predictions are immature and require further research, some progress has been achieved. For example, under the DARPA-sponsored AIM program, Bayesian analysis was conducted to predict yield strength variation. It used models within a Monte Carlo scheme to construct an a priori distribution that was then refined using only limited data. However, beyond such integrated analysis methods, models also can be applied to interpolate within sparse data sets; models, empirical rules, and heuristics can each be applied to identify suspect data and quantify/tag their uncertainty. Databases should be integrated with repositories of modeling results, rules, heuristics, and lessons learned to form a comprehensive knowledge base that will facilitate such information fusion.

researcher to find the minimum number of components that best describe a dataset, enabling much easier classification and feature extraction. Other data-mining tools include partial least squares regression, cluster identification, association analysis, and anomaly detection. These approaches are common in some fields but have yet to be widely applied to materials data.

As an example, suppose the goal is to develop a new alloy system for a specific application. The expense associated with a complete exploration of a multicomponent design space is immense and generally not affordable. Informatics provides a way to identify trends in the data that might be normally overlooked. Using data-mining methods, common characteristics can be isolated and employed to identify promising classes of materials. The power of informatics is that the data can come in many forms—for example, from experiment or from modeling—and can have a wide range of uncertainty.

To be more specific, consider the hybrid data mining and simulation technique of Fischer et al. for determining lowest-energy intermetallic structures and constructing their phase diagrams.[24] A database holds the structures and free energies of a large number of binary and ternary intermetallic systems. When the user requests the phase diagram for a binary system not in the database, the software first guesses which structures the alloy could form by applying statistical methods to the database, then tests and refines those guesses by a series of ab initio calculations. Fischer et al. estimate that an unknown binary phase diagram, including all intermetallic crystal structures and lattice spacings, could be generated in this way by using just 20 ab initio calculations. Moving up in scale, it is not hard to imagine a mesoscale structure formation and evolutions models (such as the phase field method) employing a similar approach to automatically access thermodynamics data and the results of ab initio calculations.[25,26] Indeed, this general approach may be widely applicable in linking models across scales.

For the foreseeable future, the development of ICME computational models will require a specialized capability and a labor-intensive approach requiring an "expert." A good example of this is CALPHAD, which needs experts or those experienced in the "art" to develop data assessments and assemble databases. It is an iterative process, and many of the more commonly used databases for alloys (such as Ni superalloys and steels) have been in development for up to 20 years. Within the ICME framework, work on better, more efficient ways to manage databases

[24]C. Fischer, K. Tibbets, D. Morgan, and G. Ceder, "Predicting crystal structure by merging data mining with quantum mechanics," *Nature Materials* 5 (2006): 641-646.

[25]V. Vaithyanathan, C. Wolverton, and L.Q. Chen,. "Multiscale modeling of θ′ precipitation in Al-Cu binary alloys," *Acta Materialia* 52 (2004): 2973-2987.

[26]V. Vaithyanathan, C. Wolverton, and L.Q. Chen, "Multiscale modeling of precipitate microstructure evolution," *Physical Review Letters* 88(12) (2002).

and construct them in a more semiautomated way would be an important step forward.

Materials informatics is in its earliest stages of development. Much work remains before it will be developed sufficiently to be widely applicable in materials engineering; it requires creating a new and robust set of tools easily available to the materials engineer. It holds great promise, however, and could be a critical part of an ICME process.

INTEGRATION TOOLS: THE TECHNOLOGICAL "I" IN ICME

Technical tools for integrating materials knowledge are of obvious importance for ICME. Integration tools are the glue that binds software applications and databases into an integrated, cohesive, systemwide design tool that can be used by many contributors to the design effort. For ICME, these contributors might include materials researchers, materials design engineers, product designers, engineering design analysts, manufacturing analysts, purchasing agents, suppliers, and, possibly, quality control and customer support personnel. Integration tools are required for three tasks:

- *Linking information from different sources and different knowledge domains.* This information could be in the form of computational models or empirical relationships derived from experimental data.
- *Networking and collaborative development.* This would be a helpful technical tool for solving some of the cultural and organizational problems facing ICME, which will be described in Chapter 4.
- *Optimization.* This might be optimization of a product, a manufacturing process, or a material. It would allow materials engineers to fully engage in the computational engineering IPD process described in Box 2.1.

Integration is viewed differently in each of the communities expected to contribute to the growth of ICME. Graphical representations representing the viewpoints of three of those communities are shown in Figures 3-6 to 3-8. Figure 3-6 shows a typical multiscale figure, with the timescale and the length scale important for the description of various systems. Much of the work that could be deemed computational materials science entails performing calculations in each of these regimes and then, by passing information from one regime-specific tool to another, linking the phenomena across the scales. While this concept is often useful for defining a modeling strategy, its importance is sometimes overemphasized. Developing and linking models across length scales is not required for a workable ICME tool set. Rather, ICME practitioners develop models as an engineering activity that

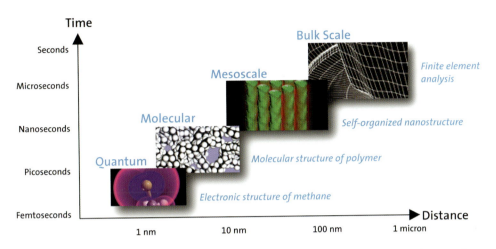

FIGURE 3-6 Multiscale modeling, a construct used to illustrate the interdependence and connections between mechanisms acting at different length scales and timescales. SOURCE: Michael Doyle, Accelrys, "Integration of computational materials science and engineering methods," Presentation to the committee on March 13, 2007. Available at http://www7.nationalacademies.org/nmab/CICME_Mtg_Presentations.html. Accessed February 2008.

requires an initial expert assessment to get proper matching between the problem being attacked and the length scales that must be considered.

Figure 3-7 (and to a large degree Figure 3.5) shows the integration problem from the viewpoint of a metallurgist. Viewpoints exist as well for ceramics, polymers, and other materials systems. Knowledge from disparate sources and domains (for example, thermodynamic models, models for simulating manufacturing processes, microstructural evolution models, and property models) is required to fully assess the influence of the manufacturing process on the properties of the materials that make up a manufactured product. Simulations of manufacturing processes must be integrated with computational models for phase equilibria, microstructural evolution, and property prediction. An important notion here is that properties of an engineering product "compete" and thus must be balanced in its design. The complexity of this optimization problem dictates that a computational approach is required. Missing from the traditional metallurgist's perspective are the direct outputs to product development performance analysis and optimization.

To be effective, ICME must address issues that are encountered in both of these integration domains and many more. In doing so, it will integrate these disparate fields into a holistic system allowing optimization and collaboration. Integration tools are thus the backbone of ICME. Depending on specific motivations, incentives, and requirements, they may be used in a proprietary setting (such as described

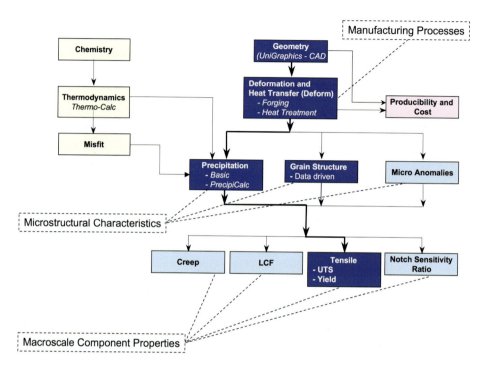

FIGURE 3-7 A metallurgist's view of the integration problem represented by ICME for a nickel-based superalloy. SOURCE: Adapted from Leo Christodolou, DARPA, "Accelerated insertion of materials," Presentation to the committee on November, 20, 2006. Available at http://www7.nationalacademies. org/nmab/CICME_Mtg_Presentations.html. Accessed February 2008.

in the Ford virtual aluminum castings example in Chapter 2), in a collaborative but limited partnership setting (such as that described for the P&W AIM program, also in Chapter 2), or in an open, collaborative setting.

Commercial Integration Tools

Commercial integration software tools are available that are designed to link a variety of disparate software applications into an integrated package, which can then be used to optimize some underlying process. As a result of these efforts, de facto standards are emerging for "wrapping" models, running parallel parametric simulations, applying sensitivity analysis, and reducing the complexity (order) of systems. Such companies market and apply systems integration tools that will solve specific engineering problems, tools for interoperability across organizations, and, in some cases, tools for education.

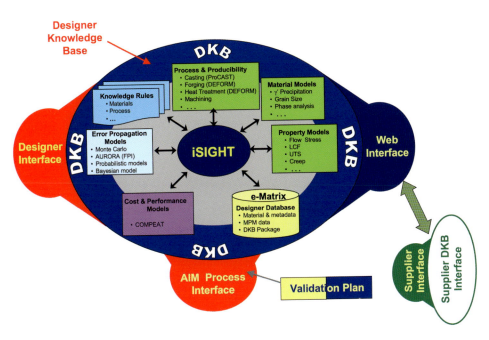

FIGURE 3-8 Models and experiments flow from AIM integration architecture. SOURCE: DARPA and AFRL Accelerated Insertion of Materials program.

Simulation data managers (SDMs) such as iSIGHT/FIPER and CenterLink are Web-based tools that do the following:[27,28]

- Provide standards-based integration environments to link applications;
- Send data securely across network connections;
- Run applications code on computer resources that might be local or remote and that consist of heterogeneous hardware platforms;
- Use system resource or job execution queue managers such as load sharing facility (LSF);

[27]Brett Malone, Phoenix, "Phoenix integration," Presentation to the committee on March 13, 2007. Available at http://www7.nationalacademies.org/nmab/CICME_Mtg_Presentations.html. Accessed February 2008.

[28]Alex Van der Velden, Engineous, "Use of process integration and design optimization tools for product design incorporating materials as a design variable," Presentation to the committee on March 14, 2007. Available at http://www7.nationalacademies.org/nmab/CICME_Mtg_Presentations.html. Accessed February 2008.

- Save design parameters and results in a database;
- Provide database mining capabilities;
- Enable three-dimensional surface design visualization;
- Provide response surface approximation for experimental data; and
- Measure and track uncertainty and contributions for given design parameters.

These and other integration tools are widely used for IPD, but they have almost no presence in the materials engineering community. That said, they have been successfully used in pilot ICME demonstration projects.[29,30,31,32] In the Defense Advanced Research Projects Agency's (DARPA's) Accelerated Insertion of Materials (AIM) program, a commercial SDM called iSIGHT, from the company Engineous Software, was used to link computer-aided design (CAD) forging process modeling, models for heat treatment, microstructural evolution models, property predictions, and structural analysis applications into a seamless work flow called a designer knowledge base. This designer knowledge base, depicted in Figure 3-8, effectively integrated quantitative information from a wide variety of sources and models. Design data and experimental results were stored in a common database. From this demonstration, the committee concludes that state-of-the-art commercial integration tools are available for ICME and ready for widespread application, identifying and solving the unique problems that will arise as the discipline matures.

For integration tools, common interface standards are highly desirable so that application engineers do not have to rewrap applications many times for different uses. The development of standards and nomenclatures, or taxonomies, should be done in conjunction with model and software developers and vendors and not in isolation. NIST has developed a wrapping standard that is available in the commercial SDM applications FIPER.[33] Other interface formats are emerging from

---

[29]Leo Christodolou, DARPA, "Accelerated insertion of materials," Presentation to the committee on November, 20, 2006. Available at http://www7.nationalacademies.org/nmab/CICME_Mtg_Presentations. html. Accessed February 2008.

[30]Daniel G. Backman, Daniel Y. Wei, Deborah D. Whitis, Matthew B. Buczek, Peter M. Finnigan, and Dongming Gao, "ICME at GE: Accelerating the insertion of new materials and processes," *JOM* November 2006: 36-41.

[31]Dennis Dimiduk, United States Air Force, "Towards full-life systems engineering of structural metal," Presentation to the committee on May 30, 2007. Available at http://www7.nationalacademies. org/nmab/CICME_Mtg_Presentations.html. Accessed February 2008.

[32]Alex Van der Velden, Engineous, "Use of process integration and design optimization tools for product design incorporating materials as a design variable," Presentation to the committee on March 14, 2007. Available at http://www7.nationalacademies.org/nmab/CICME_Mtg_Presentations.html. Accessed February 2008.

[33]Ibid.

UGS, MSC, and Dassault.[34] Additionally, the International Organization for Standardization (ISO) has promoted the Standard for the Exchange of Product Model Data (STEP), ISO 10303, as a comprehensive way to represent and exchange digital product information. However, application developers are often reluctant to create open interfaces to their applications. Open access is often viewed from a software developer's point of view as a risk to its intellectual property. To the extent that code vendors and authors are willing to cooperate in the creation of open standard interfaces to their applications and data, the general community would benefit; to encourage them to do so would require incentives from major government agencies and industrial consortia.

SDM environments provide the ability to securely transport data to local or remote computing resources and to execute a wide variety of applications on those resources. Once that has been done, the SDM takes computed results and stores them in a simulation database. SDM environments also enable collaboration among multiple research groups. Execution can be monitored by these groups, with all parties having limited or full access to the data or control of application execution. The groups can be inside or outside firewalls.

Optimization is an important objective for ICME. Once applications are linked into a common framework, the next logical step is to perform multidisciplinary, systemwide design and optimization. Design trade-offs can be made, and the resulting behavior can be propagated throughout the entire design work flow to obtain globally optimal solutions. Although materials computations are not currently integrated into the multidisciplinary optimization (MDO), the ICME-enabled desired future state would allow material and manufacturing process optimization trade-offs that could also be propagated throughout the entire design work flow. A single analysis of all of the linked application modules could be executed, or design studies could be conducted to access trade-offs. An ICME-enabled MDO system could also be used to bring the systemwide design to an optimal global design point or it could be run to simply assess reliability.

ICME Cyberinfrastructure

For many communities the World Wide Web serves as a platform for sharing information in the form of models and data. The term "cyberinfrastructure" refers to a relatively new infrastructure that according to an NSF report[35] is "based upon

---

[34]Nuno Rebelo, Simulia, "CAE: Past, present, and future," Presentation to the committee on May 30, 2007. Available at http://www7.nationalacademies.org/nmab/CICME_Mtg_Presentations.html. Accessed February 2008.

[35]For more information, see the *Report of the National Science Foundation Blue-Ribbon Advisory Panel on Cyberinfrastructure.* Available at http://www.nsf.gov/od/oci/reports/atkins.pdf. Accessed February 2008.

distributed computer, information and communication technology." Such an infra-structure is as essential to the knowledge industry as is the physical infrastructure of roads, bridges, and the like to the industrial economy. In 2003, the NSF Blue Ribbon Advisory Panel on Cyberinfrastructure envisioned "the creation of thousands of overlapping field and project collaboratories or grid communities, customized at the application layer but extensively sharing a common Cyberinfrastructure."[36] Important elements of the cyberinfrastructure described in this report included grids of computational facilities; comprehensive libraries of digital objects, including programs and literature; multidisciplinary, well-curated, federated collections of scientific data, online instruments, and sensor arrays; convenient software toolkits for resource discovery, modeling, and visualization; and the ability to collaborate with physically distributed teams of people using these capabilities. The report identified this as an important opportunity for NSF and stressed the importance of acting quickly and the risks of failing to do so. The risks include lack of coordination, which could lead to adoption of irreconcilable formats for information; failure to archive and curate data that have been collected at great expense and may be easily lost; barriers that can inadvertently arise between disciplines if isolated and incompatible tools and structures are used; waste of time and talent in developing tools that may have shortened life spans due to the above-mentioned lack of coordination and failure to incorporate a consistent computer science perspective; and, finally, insufficient attention to resolving cultural barriers to adopting new tools, which may also result in failure. The committee proposes the following definition for the term "ICME cyberinfrastructure:"

> The Internet-based collaborative materials science and engineering research and development environments that support advanced data acquisition, data and model storage, data and model management, data and model mining, data and model visualization, and other computing and information processing services required to develop an integrated computational materials engineering capability.

A key element of the ICME cyberinfrastructure will be individual collaborative ICME Web sites and information repositories that are established for specific purposes by a variety of organizations but linked in some fashion to a broader network that represents the ICME cyberinfrastructure. The DARPA AIM Designer Knowledge Base, using iSIGHT, the Internet, and a geographically dispersed team, represents the only known example of an ICME–Web collaboration. Although collaborative Web sites in materials science and engineering are relatively rare, there are some. One example is nanoHub, the Web-based resource for research,

---

[36]Ibid., p. 7.

education, and collaboration in nanotechnology, developed at Purdue University and funded by the NSF Network for Computational Nanotechnology.[37] It is reportedly used by thousands of researchers from over 180 countries. Important elements of collaborative sites are security, networking capability, and, in some cases, grid computing. Since the technology surrounding collaborative Web sites and informatics is rapidly evolving and quite new in the case of materials science and engineering, it should be realized that there may be some redundancies and that some ICME Web sites and informatics efforts probably will fail eventually. If ICME develops soon and with substantial coordination, these redundancies and failed efforts will be minimized.

The goal of a balanced, well-designed ICME cyberinfrastructure is to give scientists and engineers the means to do a number of things:

- Link applications codes—for example, UniGraphics, PrecipiCalc, and ANSYS;
- Develop models that accurately predict multiscale material behaviors;
- Store and retrieve analytical and experimental data in common databases;
- Provide a repository for material models;
- Execute computational code anywhere computational resources are available;
- Visualize large-scale data;
- Enable local or geographically disperse collaborative research; and
- Measure the uncertainty in a given design and the contributions of individual design parameters or sources.

The desired future state is one in which a government-sponsored cyberinfrastructure composed of a variety of special-purpose Web sites is widely used for collaboration between researchers here and abroad. It will be routinely used to develop materials models for computer-aided engineering (CAE) analysis of new products, linked with manufacturing simulations. The development and maintenance of advanced materials models that are sensitive to manufacturing history will, for the foreseeable future, be accomplished by specialists from industry, small business, or academia. The materials models will be used to optimize product design and the manufacturing process and to develop new materials. Because these are collaborative tools, access to and control of vital data are crucial to all members of a research consortium and should not be hindered by security considerations. Minimizing redundant activities is a key side benefit of coordinated development programs such as those devoted to solving engineering challenge problems like the

---

[37]For more information, see http://www.nanohub.org. Accessed March 2008.

ones described in Chapter 2 and will be expanded on in Chapter 4. Whether the ICME cyberinfrastructure is being put to use in the context of an effort to solve a foundational engineering problem or is part of a self-assembled activity conducted by a professional society, the extent to which it can be made openly accessible to researchers both at home and abroad will greatly influence the cost and rate of development of an ICME capability.

ICME integration tools must be made compatible with the underlying hardware and the software environment used in current design processes. Elements of computational compatibility include the following:

- Portability across the heterogeneous hardware and operating systems. Members of the IPD team (IPDT) may be located at different companies that have different computing environments.
- Interoperability with other tools and integrating software.
- Standard data and I/O formats to permit data propagation among codes.
- Efficient operation, so that tools may be used within a design optimization loop.
- Good code design practices, so that codes may be updated as models improve or computing environments change.

Standard formats for each of these data inputs and outputs are required to avoid the proliferation of formats seen in property data. As with properties, each format needs to specify uncertainty, trust, and generation method. And, formats will need to be able to evolve to meet changing needs and modeling capabilities.

Security is a vital part of the ICME collaborative integration process. Rarely can a materials designer or a developer of a material's constitutive relationship use a single code to take a design from microstructure predictions to material properties and the analysis of final product design. This systemwide analysis requires running multiple applications that might reside on different systems run by different, possibly geographically remote groups. The ability to remotely execute tools and move sensitive data in a secure manner is critical to the success of the design community. Secure access to model and data repositories is also essential. Without proper security, corporate and government security policies will impede the development of systemwide design environments.

SUMMARY

Although existing computational materials science capabilities are impressive, they have not had a significant impact on materials engineering. Moreover, computational materials science lacks the integration framework that would make it widely usable in materials engineering. Establishment of such a framework would transform the materials field. In selected instances, the existing tools have been integrated and applied in industrial settings, enabling the first demonstrations of the capabilities of ICME. Physically based models and simulation tools have progressed to the point where ICME is now feasible for certain applications, though much development and validation remain to be done if ICME is to be more broadly adopted. The widespread adoption of ICME approaches will require significant development of models, integration tools, new experimental methods, and major efforts in calibration of models for specific materials systems and validation.

The continued evolution and maturation of computational materials science tools will accelerate the ease and efficiency with which ICME tools can be implemented. To be effective in an ICME environment, all of these tools must be developed in a manner that allows their integration with other tools; this should be a priority for model developers and funding agencies. Modeling approaches that embed uncertainty are also important for advancing ICME. The advantage of improvements in computational capabilities such as parallel processing should be exploited by future CMS and ICME developers.

Although ICME tools will be used in a computational engineering environment, experimental studies and data are also critical for the development of empirical models that can be used where there are gaps in theoretical understanding and that can be used to calibrate and validate ICME models. There are several new experimental methods under development whose maturation will do much to accelerate the widespread development of ICME. These include rapid characterization methods, miniature sampling techniques, and three-dimensional materials characterization techniques. Validation experiments should be a key element of any approach to solve the engineering challenge problems that will be discussed in Chapter 4.

The creation and maintenance of dynamic and open-access repositories for data, databases, and materials taxonomies are essential. These databases can also play a role in linking models at different spatial and temporal scales. Open access databases will reduce redundant research, improve efficiency, and lower the costs of developing ICME tools.

The integration tools that are now available provide working solutions for ICME, but significant infrastructural development will be required to realize the benefits of integration. One forerunner of an ICME capability will be the establishment of curated ICME Web sites that can serve as repositories for data, databases, models for collaboration, and model development and integration. Significant government investments, similar to those awarded by the NIH to the genomics research community, will be required to create and curate the cyberinfrastructure necessary to support ICME. The extent to which this ICME cyberinfrastructure can be made open and accessible will greatly speed up the development of an ICME capability and lower its cost.

4

The Way Forward: Overcoming Cultural and Organizational Challenges

A lesson learned from the case studies considered by the committee and from other fields that have undertaken major integration efforts is that the cultural issues account for more than half of the effort required for making progress and gaining acceptance. For integrated computational materials engineering (ICME) to gain widespread acceptance, cultural shifts are required within industry, academia, and government. It was clear from several case studies that the cultural changes required to fully benefit from ICME should not be underestimated. In the sections that follow the committee identifies the primary cultural barriers faced by the communities involved in advancing and implementing ICME. In this chapter, the committee considers the cultural barriers faced by the key stakeholders in industry, government, and the materials science and engineering (MSE) community. Its recommendations address barriers to industrial acceptance of ICME within the integrated product development (IPD) process, the acceptance by government of its role as a champion and coordinator of ICME, and acceptance in the MSE community of ICME as an inherent component of its overall education, identity, profession, and practice.

CULTURAL BARRIERS TO ICME IN THE MANUFACTURING INDUSTRY

As described earlier, the IPD process has revolutionized U.S. industry. (For a discussion of the IPD process, see Box 2-1.) However, although the constraints imposed by the choice of materials for a particular product can strongly influence the design and manufacture of that product, materials are not currently part of the IPD computational-tool-based optimization process. The constraints of materials are only considered outside the IPD multidisciplinary design loop, and materials are a fixed and limiting constraint on the overall IPD process rather than a parameter that can be optimized along with other engineering parameters. The allowable list of materials may be taken as fixed or it may include a small subset of materials that are evaluated outside the optimization loop. This approach narrows the design space, resulting in suboptimal product performance.[1] Conversely, the development of the optimal materials, currently a lengthy and expensive process, does not benefit from integration of the materials and materials-manufacturing development processes into the product optimization process.

Although there have been several successful demonstrations of ICME applications within industry (see the case studies discussed in Chapter 2), as Lesson Learned 1 from that chapter indicates, ICME is still in its infancy and ICME technologies remain immature. Indeed, in the committee's judgment, gaps between the available and the ideal computational models and methods will persist for some time. While this does not preclude the development of an ICME capability for a particular material or application, it can make the capability much more difficult to achieve. Moreover, and perhaps more important, for many industrial firms the existence of ICME and its benefits are unknown. The engineering organizations that have experimented with the application of ICME have judiciously selected engineering challenges with near-term ICME opportunities. Focused sets of materials tools have been or are being integrated to address a specific engineering component and material system. However, the committee knows of no organization where ICME is fully institutionalized or represents the norm for the product development process.

Inertia in the Engineering Design Community

The relatively rapid adoption by industry of the IPD process and computationally based multidisciplinary design optimization (MDO) indicates that the integration of engineering tools and computation has become and will remain a critical element in product design and development. However, the insertion of materials into this loop to improve product design and optimization may actually be impeded by industry's prior investment in and commitment to existing integrated engineering tools and processes. The competing business forces that encourage and dissuade ICME implementation are shown in Figure 4-1. Many engineering

---

[1]The design space is the set of possible designs and design parameters that meet a specific product requirement. Exploring the design space means evaluating the various design options possible with a given technology and optimizing the chosen option with respect to specific constraints such as power or cost.

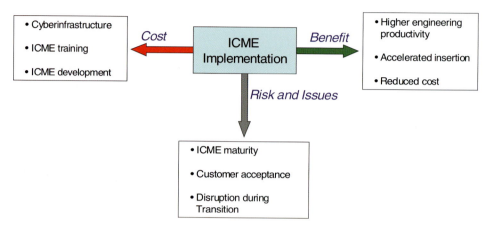

FIGURE 4-1 Competing business factors that affect the decision to implement ICME technical maturity and validation.

organizations, particularly at large companies, have invested considerable human resources and capital in the codification of their engineering practices and product development processes and timelines. Engineering organizations capture best practices, provide discipline in the design and execution of engineering programs, and build consistent, validated, and approved computational methods. While these formal engineering practices and product development processes enhance the engineering product, they also retard the adoption of progressive, promising new directions such as ICME.

As depicted in Figure 4-1, before ICME reaches full maturity and acceptance and before it is validated, engineering managers will take a cautious stance to its implementation where its development costs are high and where they might be subject to liability risks when ICME-based recommended practices do not agree with previously accepted materials engineering practices.

For many industries, the design engineer or the production engineer is the leader of the IPD and, in a sense, the primary customer for the materials engineer. The design engineer depends on materials engineers for material selection advice, material property data or constitutive relations (that is, the relation between applied stresses or forces and strains or deformations) for design analyses, and materials evaluation of tested components or components that have already been in use. In general, design engineers who have confidence in the existing support role of materials engineering are likely to resist any extensive, rapid change—such as might be entailed in adopting an ICME approach to materials engineering—because their design methods and inputs have withstood the test of time and gained customer acceptance. For example, a design engineer probably will not want to spend time reconciling the product life predictions from proven, data-driven analyses with predictions from ICME when he/she feels confident that existing methods work and when there is skepticism about the value of materials computational methods and tools. Also, design engineers focus on the demands of their own discipline and generally have less understanding of and interest in material mechanisms and models, and they may be constrained by formal design practices and regulatory requirements that cannot be modified without rigorous validation and approval by a chief engineer or regulatory agency.

The industrial product development community must have confidence that ICME has sufficient maturity and validity to provide, at acceptable risk levels, tangible offsetting benefits in terms of product improvement, cost savings, and/or reductions in the development cycle time. Given the option to invest in the new and "unproven" (to their way of thinking) capability provided by ICME, many engineering and R&D managers will choose the traditional route to a new product in the absence of some incentive.

Inertia in the Industrial Materials Engineering Community

Although the role of the materials engineering discipline's support of IPD varies from industry to industry and firm to firm, it can undermine confidence in ICME and prevent its adoption. Given the current state of MSE curricula and computational materials tools, experienced materials engineers are often ill equipped to develop or use ICME tools to perform their job. They may be skeptical of ICME capability and will insist on acquiring their own data to support their engineering decisions. A further cultural barrier is that materials engineering managers may resist adopting radically new engineering processes that have not demonstrated the ability to deliver the required materials technologies within established product development and investment cycles. For the foreseeable future, ICME tool sets will be developed by specialists, either in industry or academia. In large and medium companies this role may fall to either in-house materials researchers or materials engineers with specialized training. Materials engineering managers, cognizant of a gap in ICME skills, may avoid large-scale, rapid ICME deployment, especially if they believe that ICME requires the addition of new employees with a higher level of education.

In addition, managers who have traditionally imposed restrictions on the external disclosure of materials information will need to redefine what information is truly sensitive and what information can safely be disseminated among the community for mutual benefit. This will be more readily accomplished in industries where specific materials do not represent a strategic advantage—for example, the automotive, consumer product, and packaging industries.

In industries where the materials of construction are not themselves proprietary, the primary resistance to sharing data is the high cost to develop a material. Mechanical engineering managers do not worry about who among their competitors is using a particular finite-element analysis (FEA) software package, and—similarly—progressive materials engineering managers will take the same view of ICME-enabling infrastructures. Rather they will emphasize the benefits of skillful use of ICME tools. Their support for open-access databases will accelerate the further development and adoption of ICME.

Materials engineering leaders and decision makers must also identify a pathway by which to transition gracefully from current practice to ICME without disrupting the supply of materials to design, manufacturing, and product support operations. A successful pathway must account for the needs imposed by ICME on the skills and training of materials engineers within an organization, the handling of proprietary information and information controlled by regulation, the expansion of computational infrastructure, and the risks perceived by customers.

Inertia in the Manufacturing Engineering Community

Resistance to adopting ICME may also come from outside the design and materials communities—namely, from manufacturing engineers. One of the primary objectives of ICME is the integration of manufacturing and product design into a holistic computational system that includes the materials developer, leading to an increased role for manufacturing engineers in providing new and quantitative outputs to the materials and product development process. In most industries, the main focus of computer-aided engineering (CAE) for manufacturing is to make sure the part being designed can be manufactured. In contrast, the main focus of design CAE is to design the "perfect" part, without considering manufacturing constraints or variability. Even in IPD, design CAE analysts often work in isolation from manufacturing simulation, and the integration of manufacturing CAE and design CAE is rarely considered. Integrating design, materials, and manufacturing engineering using computational tools and methods will require overcoming organizational chimneys, rising above the resistance to change established processes, and adjusting product and process development cycles to provide or receive proper and interconnecting inputs and outputs. A significant barrier to achieving this kind of integration is that manufacturing CAE is often conducted by suppliers, and in general there is no paradigm for passing manufacturing CAE information to original equipment manufacturers (OEMs). This results in barriers associated with purchasing policies and concerns about proprietary information.

Overcoming Inertia in the Manufacturing Industry

Even if there were no other barriers to the implementation of ICME, many engineering organizations would opt to either delay it or phase it in owing to the uncertainty surrounding its completeness and the lack of maturity for the full suite of ICME models and tools. Prudent organizations will critically identify, assess, and prioritize those applications where ICME methods add value to their engineering function and then judge which are feasible with the existing state of the art. Most firms will judge the technology readiness by comparing ICME's results with the results of existing, proven, data-driven methods. Organizations will encounter lower introduction barriers if they implement ICME gradually and initially learn how to use it to supplement existing methods or provide results that fuse traditional and ICME methods. Conversely, once ICME achieves greater acceptance, current data-driven methods can be applied to establish empirical relationships that can substitute for missing physically based models in an ICME system. Regardless of the ICME implementation strategy, implementers must emphasize validation and uncertainty measurement to gauge and demonstrate progress in the maturation of ICME and provide a sound basis for its application.

It is important to recognize that such cultural barriers can be and have been overcome in certain sectors where ICME's value is recognized. One prominent example of such a cultural shift occurred in the auto industry. Government standards require crash testing of new vehicle designs, which has led to the costly and time-consuming building of vehicle prototypes and their crash testing during the design process. A complicating factor is that the response of prototypes is not necessarily the same as that of production cars. Thus a sufficiently accurate computational model offers large payoffs. For example, in the mid-1980s, a standard vehicle development testing program would have required approximately 30 crash tests per year for 5 years, each at a cost of approximately $250,000. For the Ford Mondeo program, finite element modeling enabled a reduction in test costs of 30 percent, or $12.5 million, over 5 years.[2] Since that time, despite the imposition of more stringent safety requirements, the number of crash tests has been cut in half through the use of CAE. Recognizing the utility and validity of computational crash testing, automakers have shifted their cultures and rely increasingly on CAE to cost-effectively engineer safe vehicles.[3,4]

A critical challenge is providing the human and financial resources required to fully develop a firm's ICME capability. In the near term, overcoming this challenge

---

[2]David Bensen, University of California, San Diego (UCSD), "Integrating FEM with materials models for crash tests," Presentation to the committee on March 13, 2007. Available at http://www7.nationalacademies.org/nmab/CICME_Mtg_Presentations.html. Accessed February 2008.

[3]P. Prasad, Ford Motor Co., private communication (2007).

[4]David Bensen, UCSD, "Integrating FEM with materials models for crash tests," Presentation to the committee on March 13, 2007. Available at http://www7.nationalacademies.org/nmab/CICME_Mtg_Presentations.html. Accessed February 2008.

will require investment in ICME development across different product development cycles.

The term "industry" implies a monolithic entity with common interests and motivations. This is clearly not the case. So the ability of "industry" to organize a coordinated development of ICME is limited, although the establishment of consortia of companies with common interests within industrial segments is one way to foster development of ICME, particularly with augmented governmental support. Such consortia could be funded by industry alone or jointly by industry and government agencies. To focus resources and demonstrate the capability of ICME, an appropriate foundational engineering problem could be selected, as described in Lesson Learned 10 in Chapter 2. Generally a foundational engineering problem would consist of a manufacturing process or set of processes; a material system; and an application or set of applications that define the critical properties and geometries. Examples of such foundational engineering problems are the ICME capabilities described in Chapter 2 for forged nickel-based superalloy turbine disks (aerospace sector) and the cast aluminum power train components (automobile sector).

As mentioned in Chapter 1, the committee envisions that ICME could impact a range of industries. Some possible foundational engineering problems are these:

- High-dielectric materials and processes for improving the performance of microelectronic devices,
- Low-cost organics for robotics sensors,
- Thermal protection materials for hypersonic vehicle surfaces,
- Catalysts for optimizing the performance of hydrogen-fueled systems,
- Reliable and rapid recertification of components in aging structures,
- Materials for ballistic and blast survivability of ship hulls,
- Thermoplastic injection-molded materials for automotive structures,
- Materials and electrochemical processes for advanced batteries,
- Nanoparticles for magnetic storage devices, and
- Composite or advanced metallic materials for aeroengine components.

A good model of a consortium in this area is the U.S. Automotive Materials Partnership (USAMP), an ICME project in the U.S. auto industry. Ford, General Motors, and Chrysler have selected "Magnesium for Body Applications" as a foundational engineering problem for the consortium, and together they are developing an ICME infrastructure and knowledge base for these materials and the manufacturing processes that are used to fabricate engineering components from them. This 5-year international program was initiated in 2007 and is jointly sponsored with the U.S. Department of Energy (DOE), the China Ministry of Science and Technology, and Natural Resources Canada.[5] The approximate funding over 5 years is between $6 million and $7 million. It involves participation from researchers at Chrysler, Ford, and General Motors, and more than 15 universities and government laboratories. This level of funding is acknowledged to be small relative to the needs for this particular foundational engineering problem, but it is an important first step. Similar consortia would appear to be appropriate in aircraft engines, airframes, and electronic materials to name a few obvious areas.

Despite the clear vision and persuasive logic of ICME, the cultural challenges outlined above will impede the widespread implementation of ICME in industry. There is a reluctance to fix a process that is generally not thought to be broken, particularly one that was expensive to build. In summary, the committee concludes as follows:

> **Because of ICME's relative immaturity, the remaining computational gaps, and the potential for ICME to disrupt current IPD processes, the industrial product development community is skeptical of or unaware of the benefits of ICME. Sustained funding across multiple product development cycles will be required to advance ICME and build confidence in it to the point where it is fully accepted.**

> **There are two main cultural challenges impeding the widespread industrial adoption of ICME:**

> - **IPD engineers are not aware of ICME (including tools and suppliers) and ICME has not been accepted into the IPD process because its value in specific problems of interest to industry has not been proven.**
> - **Resources for things such as R&D investments and personnel must be committed to an ICME project for more than 1 year, which is longer than the typical 1-year investment cycle and outside the typical product-oriented R&D cycles.**

> **Establishing consortia within industrial segments having common interests offers a means to foster development of ICME, build awareness and confidence, and augment governmental support. It may be that government will have to offer incentives to enable such consortia and collaborations, and the terms of the consortia may need to explicitly state that the results will be open to the wider ICME community, as appropriate.**

---

[5]Joseph Carpenter, DOE, "DOE's work on extruded long-fiber-reinforced polymer-matrix composites," Presentation to the committee on March 13, 2007. Available at http://www7.nationalacademies.org/nmab/CICME_Mtg_Presentations.html. Accessed February 2008.

CULTURAL BARRIERS TO ICME IN THE MSE COMMUNITY

Advancing ICME will also require adaptation by the materials community in the United States and abroad. There must be a shift in the MSE community's mind-set so that the materials discipline becomes focused on problem solving alongside the scientific endeavor—that is, making materials engineering a more quantitative discipline and positioning MSE as an integral part of the engineering process. In addition, the MSE acceptance of ICME can be accomplished by shifts in education and research and by advocacy and communication as well as support and encouragement by government agencies and industry.

The changes needed in the MSE community can be assessed by looking at the traditional roles of MSE practitioners and how they might have to change if ICME is to proceed successfully.

Need for Change in the Roles of MSE Professionals

An environment characterized by multidisciplinary, collaborative teamwork is integral to the future success of ICME. The modern materials expert must work across disciplinary boundaries in multidisciplinary teams throughout the design process, sharing information and communicating outside the original company, country, or discipline. This requires an ability to work with others and collaborate across disciplinary boundaries, a recognition that materials are not the only design parameter or constraint, and a mind-set for sharing data and information. While the MSE community will be a primary contributor to the development of ICME, experts from a variety of other fields, including engineering, physics, mechanics, information sciences, systems engineering, mathematics, and computer sciences, will also play important roles in product design and manufacturing and in MSE research.

Product Design and Manufacturing

Materials engineers have adopted an approach to their profession that often focuses on acquiring material property data sets from which they calibrate average (nominal) materials properties, sometimes with statistical information, and, in collaboration with CAE analysts, reduce these data to conventional constitutive models for use in performance CAE. But for ICME to be truly successful, materials engineers will need to think of ICME as a key enabler capable of providing quantitative information in the product development process. This will be a significant cultural shift for materials engineers.

Materials designers, by contrast, often optimize an alloy or process for a single property without regard to other design constraints—for example, other properties, product geometries, and so on—or manufacturing constraints. Alloy development projects tend to be Edisonian and experimental rather than driven by model predictions. While designers make use of the relationships between structure, underlying physical mechanisms, and macroscopic properties, they do not use quantitative analytical or computational tools. For ICME to progress, materials designers will need to look at the material development process as an integrated computational process involving optimization of multiple attributes (microstructure, properties, geometries, and manufacturing cost) and its role in overall product design.

Materials scientists traditionally work as single investigators to gain narrow but deep technical knowledge and insights, generally not part of a larger integrated effort. Individual materials scientists can, however, provide other ICME practitioners with important inputs, and they can also be beneficiaries of inputs from those same practitioners. For such benefits to accrue, the model that focuses on individual investigators gaining deep scientific insight must be expanded to include the integration of these insights into quantitative tools that combine collective insights from many different experts. A concomitant increase in the role of materials knowledge in computational system analysis is paramount, leading to quantitative tools used by product engineers and materials engineers working together in integrated teams.

MSE Research

Within the MSE community there is a tension between the materials-science-based, fundamental efforts and the materials-engineering-based, applied efforts. ICME provides an important linkage between these two activities and the two categories of experts—scientists and engineers—and can have a profoundly important synergistic benefit. MSE academic research is playing a key role in developing ICME computational tools and experimental data sets. Although academic research is typically fundamental and not geared toward immediate industrial application, some transformational steps could allow the research to more readily translate into industrial needs by means of ICME. Protocols could be developed to facilitate a viable and easy translation. As discussed above, small businesses can play a vital role in this translation. For example, in an IPD process, a new computational modeling capability developed in academia could be readily incorporated into an IPD framework if there was a known interface format for the model or for the experimental data passing through the model. Such standards or user interfaces could be part of the IPD framework, enabling and encouraging researchers to formulate their model interfaces for easier incorporation into IPD processes, speeding up their use by others in a win-win situation for the researcher and for the ICME/IPD users. In addition, various forms of experimental data could have specific reporting formats depending on data type and could be deposited and shared, enabling the more ready use of data in model validation, data mining, and materials informatics. Chapter 3 discusses in greater depth the database needs for ICME. Such an approach would require establishing database depositories and shifting the mind-set of the researcher to include preparation of data in a standard form for inclusion on a curated, shared database. Finally, ICME capabilities can enrich fundamental materials research by providing new insights at the interfaces between different subfields and by providing a focus that enables identifying important new or underexplored areas of research.

Materials professional societies have a unique opportunity to foster ICME and make the MSE culture more collaborative and fertile for it. This support can come from assisting the materials research community in the development of standards and taxonomies, encouraging and providing incentives for open access to research results, and providing ICME collaborative Web sites, which can serve as repositories for high-quality data and models that are important parts of these research results. By hosting the ICME cyberinfrastructures described in Chapter 3, materials professional societies can enable the materials community to contribute to and in some cases to self-organize in the establishment of a broad ICME infrastructure. Professional societies can also communicate the results and successes of ICME developments in their publications and programming. The ICME efforts of The Minerals, Metals & Materials Society (TMS) stand out in this regard. TMS has established an ICME coordination committee and an ICME Web site and is actively programming and publishing in this area.[6]

In summary, the committee concludes as follows:

For ICME to succeed, it must be embraced as a discipline by the international materials science and engineering community, leading to changes in education, research, and information sharing. This would transform the role of materials science and engineering to one of uniting engineering and scientific endeavors into more holistic and integrated activities.

Education and Workforce Readiness

Implementing cultural change in the materials discipline will require the integration of ICME into the MSE curriculum if ICME is to become part of the identity of an MSE professional. With the recent reforms in engineering accreditation, the role of materials in design and the importance of computation in materials engineering undergraduate curricula are now recognized, and graduates must demonstrate the following:

---

[6]For more information, see http://materialstechnology.tms.org/icme/ICMEhome.asp. Accessed February 2007.

- An integrated understanding of the scientific and engineering principles underlying the four main elements of the field: structure, properties, processing, and performance.
- The ability to apply and integrate knowledge from each of the above four elements of the field to solve materials selection and design problems.
- The ability to utilize experimental, statistical, and computational methods consistent with the program educational objectives.

Today, practicing materials scientists and engineers solve problems by analyzing property data, examining microstructures, and drawing on their collective data and experience. For most mature engineers, this approach is consistent with their education, particularly at the baccalaureate level. Program criteria in mechanical, civil, chemical, and electrical engineering emphasize advanced mathematics to a greater degree, and curricula in those fields use relevant commercial software tools to a much greater extent than do materials curricula. Whereas a current mechanical engineering graduate often has experience with CAD and FEA software used in a broad spectrum of industrial environments, materials engineering graduates may have had only limited exposure to materials software. Certainly the introduction of ICME will challenge the current skill set of practicing materials engineers. It will also challenge the university system to find better ways to establish stronger and more coherent mathematical threads throughout the materials engineering curricula and incorporate computational methods and multidisciplinary integration more fully into tomorrow's coursework. Future university graduates equipped with the right mathematical, computational, and integration skills needed for materials engineering within design and manufacturing will fuel the maturation and application of ICME.

Outlined below are the broad principles that the committee, in its best judgment, believes must be added to the curricula to further ICME as a discipline. The committee notes that different programs will have their own ways of implementing these principles based on their current offerings and environment.

Undergraduate Education

Historically, undergraduate education in materials science and engineering has been long on the descriptive nature of materials behavior and light on quantitative analysis and computation. This is in part due to the fact that many materials phenomena, such as plastic deformation, are not easily described by differential equations with well-defined boundary conditions that can be solved by standard numerical methods. An additional challenge is the breadth of the timescales (from picoseconds to years) and length scales (from angstroms to meters) that are important to predicting materials performance. However, computational science tools are rapidly evolving in their numerical sophistication and in their predictive power. Thus preparing students for an environment where they are expected to use new computational tools in their work will mean more emphasis on the basics that underlie simulation methods as well as on computation. The challenge will be to find the correct balance so as not to lose the specialized knowledge that makes students valuable.

Broadly speaking, the incorporation of new content into a curriculum must be accomplished under the usual course load constraints. Preparing students for ICME will necessitate formal educational elements in both computation and integration. Mathematical analysis (differential equations, calculus, linear algebra, numerical techniques, programming, and so on) must remain in the curricula so the student accepts analysis as an integral part of his or her job as a materials engineer and also has the foundation for a more in-depth understanding of the methods. It may be best to include computational elements by integrating modeling and simulation throughout the curriculum, as opposed to having a few specialized computation courses. For example, a thermodynamics course could make use of standard software as a regular part of the lesson rather than a one-off event; a course on the mechanical behavior of materials might introduce finite-element calculations so the students can predict the response of a structure before a test; and so on. The key will be to ensure that the student learns that an understanding of materials and their development can be derived not only from experiments but also from simulation or from a combination of experimentation and modeling. Implementing changes like this is more challenging when computational tools are incorporated into more traditional courses, in part because of the need to convince faculty of the importance of this alternative approach. Because integration of computational tools into the undergraduate curriculum is not such an easy task, it could be facilitated by collaboration among universities. Professional societies could play a key role here. Additionally, the MSE community has a coordinating body, the University Materials Council (UMC), whose membership consists of department chairs of accredited MSE programs in the United States who could also provide leadership.

Beyond the addition of analytical and computational elements to the curricula, another challenge will be to implement an integration component into the senior year, whereby students must synthesize their knowledge across the MSE domain space in a collaborative materials team effort or, more ambitiously, integrate their materials engineering into a large, multidisciplinary engineered system in an integrated, collaborative team project. Such capstone design courses are required in many undergraduate engineering programs, including most MSE programs. The challenge is to integrate realistic materials computational tools to solve a challenging design problem. Developers of computational tools in other engineering disciplines routinely provide their software to engineering programs at a minimal cost,[7] and this permits students to become familiar with the tools that they are likely to immediately encounter in their professional careers. However, currently available free materials computational codes do not integrate materials with design in a way that would be usable for undergraduate education. If developers of materials software are slow to adopt the approach used in other areas of engineering, there will arise financial barriers to the introduction of materials tools into the MSE undergraduate curriculum.

Graduate Education

It is equally important to educate MSE graduate students in the ICME process since they are likely to be called upon to provide leadership in the development of ICME systems. At the graduate level, classes in modeling, simulation, and systems integration would enable students to specialize according to their interests as well as gain a broader knowledge of product design and manufacturing.

Graduate specializations in ICME (as distinct from computational materials science) could start with a base of courses to bring the students up to speed on the necessary computation and analysis. Additional courses could focus on specific methods, with a strong emphasis on linkage with data and other types of calculations. Educating students in modeling and computation as well as in materials informatics and their respective roles in ICME would be important for a graduate program. At the graduate level students could benefit from access to freely available research codes, and with this background they would be well prepared for ICME-based research as well as modern industrial practice. Educating systems engineers in the ICME process through graduate advanced degree programs would also further accelerate the inclusion of ICME in IPD.

Professional Development

The MSE workforce will also require professional opportunities to broaden their skill sets to include ICME, whether for direct use in engineering practice or for managing materials engineering in the workplace. In particular, introducing IPD and the ICME process and culture to MSE practitioners through workshops, short courses, and tutorials in modern MSE computation (simulation methods and tools for predicting materials structure, chemistry, physics, and properties; database mining; materials informatics; and so on) will raise awareness of the untapped potential

---

[7]David Hibbit, Abaqus, Inc., "A perspective from a commercial finite element software vendor," Presentation to the committee on May 29, 2007. Available at http://www7.nationalacademies.org/nmab/CICME_Mtg_Presentations.html. Accessed February 2008.

of ICME and how incorporating it into industrial processes can impact product design and manufacturing. As described above, materials professional societies have an important role to play in establishing an ICME development network and infrastructure and in communicating the progress and success of ICME. Materials societies also have an important role to play in including ICME in the continuing education of professionals.

The committee considers that the UMC is in a unique position to influence curricula and change the culture of MSE academic institutions and that it could take an active role in promoting ICME and the curricular changes that support improvements in the computational ability of the students who graduate from their departments. Materials professional societies can share best practices on the development and introduction of computational tools into MSE curricula. The committee also believes that materials professional societies can meet the need for workforce training by offering workshops and tutorials in ICME. Finally, the committee believes that alliances between the small businesses who are developers of software tools and MSE teaching institutions can be particularly effective for propagating ICME tools, particularly as students make the transition to engineering practice.[8]

ROLE OF SMALL BUSINESS IN ICME DEVELOPMENT

During the course of the study, the committee heard from a number of small science and engineering companies that have played a key role in developing, advocating, and maturing ICME technologies.[9] They have acted at times as the scouts for the OEMs in identifying and integrating MSE research into viable commercial products. In the course of these briefings, the committee learned that one significant concern of the end user of the products discussed is the long-term support and sustainability of the products. The committee observed a variety of business models employed in small ICME firms that may serve as models for emerging small businesses in the ICME sector. In one approach, user requirements are reviewed through consortia, focus groups, and special interest groups, with consortium members paying for a seat at the table, giving them access to the decision-making and priority-setting process for new model development. The resultant software applications are tailored to specific needs in industries that produce products ranging from pharmaceuticals to consumer goods and are licensed by the industries that

---

[8]Nuno Rebelo, Simulia, "CAE: Past, Present, and Future," Presentation to the committee on May 30, 2007. Available at http://www7.nationalacademies.org/nmab/CICME_Mtg_Presentations.html. Accessed February 2008.

[9]The committee received briefings from Accelrys (www.accelrys.com), Automation Creations (www.aciwebs.com), Engenious Systems (www.engenious.com), Materials Design (www.materialsdesign.com), Phoenix Integration (www.phoenix-int.com), and QuesTek (www.questek.com).

specified development goals.[10] Other companies are building materials and process modeling tools and offer ICME as a service by applying systems integration tools to solve customer problems.[11,12] Another well-established business model for small firms is the development of databases; an example of success is the CALculation of PHAse Diagram (CALPHAD) software and the thermodynamic databases that have become key elements of ICME.

There are several candidate business models for building and marketing databases, ranging from fully proprietary to government funded, including the following:

- Commercially available proprietary databases (Thermotech, CompuTherm, Granta, Thermo-Calc),
- Consortium (Sematech),
- Advertising-supported (MatWeb),
- Community/emergent (Wikipedia),
- Government support (National Institute for Standards and Technology, National Science Foundation, National Science Digital Library, and Center for Computational Materials Design), and
- Hybrid government/industry or government/university collaboration (NIST Solder Database[13]).

Each business model has advantages and disadvantages in its ability to meet goals for ICME, such as relevance, lack of bias, verification, innovation, efficiency, and broad availability. While any of the above models could emerge as the preferred structure(s) for ICME, each will face challenges to being accepted by the current industrial community. For U.S.-based firms, SBIR and STTR funds are generally critical for the success of these firms. That said, it must be recognized that not all the small businesses that are developing or are capable of developing ICME tools and models are U.S. firms.

---

[10]The committee notes that ICME will challenge how users want to use existing and new software tools, and this in turn will challenge how software vendors currently license their software; these issues will also need to be explored and addressed in due course.

[11]Frank Brown, Accelrys, "Scientific business intelligence," Presentation to the committee on May 30, 2007. Available at http://www7.nationalacademies.org/nmab/CICME_Mtg_Presentations.html. Accessed February 2008.

[12]Charles Kuehmann, QuesTek, "Experiences integrating theory & experiment & industrial practice/culture," Presentation to the committee on March 13, 2007. Available at http://www7.national academies.org/nmab/CICME_Mtg_Presentations.html. Accessed February 2008.

[13]Ursula Kattner, NIST, "Thermodynamic databases in support of materials," Presentation to the committee on May 29, 2007. Available at http://www7.nationalacademies.org/nmab/CICME_Mtg_Presentations.html. Accessed February 2008.

Another business approach that has been adopted[14,15] involves a small business partnering with a larger company to share the cost of developing, say, a new material model or a new computational feature that is needed to simulate their particular problem. In return, each partner company has access to the advances as soon as they have been completed.

Commercial integration software tools are available that are designed to link a variety of disparate methods into an integrated package that can then be used to optimize some multistep or multiphysics process.[16,17] Software tools such as these have frequently been developed by small to medium-sized software companies using government (for example, SBIR) funding. As a result of these efforts, de facto standards are emerging for wrapping models, running parallel parametric simulations, sensitivity analysis, and reducing the complexity (order) of systems. While these codes are becoming widely used for IPD and MDO, they are also well suited for ICME. Such companies market and apply systems integration tools that will solve specific engineering problems, tools for interoperability across organizations, and tools that are experiencing a period of sustained business growth. The rapid adoption of IPD by industry is further evidenced by the large international participation in IPD conferences; in particular, although IPD was developed in the United States, it has been widely adopted by competitors from abroad, and now licenses for this MDO integration software and attendance at IPD conferences in Japan are reported to be twice that in the United States.[18]

Based on information from a representative group of small ICME businesses, the committee concludes as follows:

Small science and engineering companies are playing a key role in developing, advocating, and maturing ICME technologies. They act as the scouts for the OEMs in identifying and integrating MSE research into viable commercial products. These innovative firms develop just the kinds of technologies

---

[14]David Hibbit, Abaqus, Inc., "A perspective from a commercial finite element software vendor," Presentation to the committee on May 29, 2007. Available at http://www7.nationalacademies.org/nmab/CICME_Mtg_Presentations.html. Accessed February 2008.

[15]Nuno Rebelo, Simulia, "CAE: Past, present, and future," Presentation to the committee on May 30, 2007. Available at http://www7.nationalacademies.org/nmab/CICME_Mtg_presentations.html. Accessed February 2008.

[16]Brett Malone, Phoenix, "Phoenix integration," Presentation to the committee on March 13, 2007. Available at http://www7.nationalacademies.org/nmab/CICME_Mtg_Presentations.html. Accessed February 2008.

[17]Alex Van der Velden, Engineous, "Use of process integration and design optimization tools for product design incorporating materials as a design variable," Presentation to the committee on March 14, 2007. Available at http://www7.nationalacademies.org/nmab/CICME_Mtg_Presentations.html. Accessed February 2008.

[18]Ibid.

and human expertise that are required to mature ICME and that the SBIR and STTR programs were designed to support.

PROPOSED APPROACH TO ICME FOR THE GOVERNMENT

As described above, significant effort will need to be expended in industry and within the MSE community to support the widespread development and adoption of ICME. But there is another key stakeholder. The federal government supports materials research with two goals in mind: (1) to further the fundamental understanding of materials science and materials engineering and (2) to promote the adoption of materials technologies in industries of national importance—that is, in priority areas such as national defense, the energy sector, automobiles, aerospace, and health care. Government support of materials engineering R&D often focuses on technologies that can be successfully matured and implemented in a specific product. There are numerous examples of successful research translating into new capabilities that underpin the competitiveness of the country and its economic and national security.

The committee believes that government agencies play a critical role in championing, developing, and implementing ICME in the United States. Bringing ICME from its infancy today and turning it into a mature, widely adopted element of the materials profession in academia and in the industries that rely on materials engineering will require the resources, organizational capability, and the collective long-range vision that are best supported by government. The committee is convinced that just as government has undertaken coordinated activities that supported the development of nanotechnology from its nascent state 10 years ago; the mapping of the human genome; and the development of computing, networking, and software technologies, it can clearly also play a crucial role in the development and validation of an ICME methodology calling on the capabilities of its research agencies. Although not strictly a cultural challenge, insufficient coordinated government support for ICME would constitute a nontechnical barrier to ICME's emergence as a mature discipline. Government has the opportunity to underpin the cultural changes described in this chapter and the technical challenges described in Chapter 3. Government agencies are uniquely able to organize, advocate, and sustain the long-range programs needed to make ICME a reality. Absent a significant level of government organization and funding progress, development of ICME will take many decades and might never fully materialize.

So far federal research investment in tools for computational materials science has focused mainly on supporting individuals or small teams to develop tools for specific materials technologies. There have been successful government-supported programs for ICME-related activities—for example, the Advanced Insertion of

Materials (AIM) program supported by the Defense Advanced Research Projects Agency (DARPA) and the Advanced Simulation and Computing (ASC) program supported by the National Nuclear Security Administration (NNSA)—and new efforts such as DOE support of ICME development in the automobile sector. The success of these efforts has made ICME ready today to benefit from a systematic, sustained program to develop the tools and infrastructures needed for ICME. The right government action now can promote the development of the technical tools required and the sharing of the data and models developed by MSE R&D with the engineering community. Integrating the disjointed activities of today will require a change in philosophy and a coordinated, substantial, long-term effort but will result in less duplication of effort and less wasting of precious R&D resources. If properly captured and designed, much of the current government R&D portfolio could be integrated into ICME efforts, with potentially a very modest amount of new funding. The committee speculates that this could be accomplished by targeting existing research funds and eliminating redundant research. Incremental funding would clearly be required for activities such as material informatics, cyberinfrastructure, and database development and curation, which are not currently being funded in any meaningful way within the materials community.

As was discussed in Chapter 3, while some of the gaps in the ICME tool set have been bridged, the full set of tools for ICME remains to be developed. The time is ripe to develop concerted, coordinated efforts to calibrate existing models for particular materials systems and applications and integrate them into usable integrated models that include manufacturing, materials, and design inputs and outputs. Where modeling gaps exist, the experience summarized in Chapter 2 has shown they can generally be filled by the empirical relationships developed for the particular material system under consideration. The experience to date has shown that making progress in ICME capabilities is generally an organizational challenge and that the rate of progress in ICME capability will be directly related to the degree of organization and level of funds directed at the problem.

ICME is an emerging discipline. In most cases it is a precompetitive activity, although, as with nanotechnology, niche commercial applications are likely to emerge. Sustained funding as part of longer-term R&D efforts will be required given the need for the discipline's maturation and the need to train researchers who can effectively contribute and educate engineers who are critical to its implementation. Firms can also reap financial benefit from ICME capabilities, suggesting to the committee, as mentioned above, that a useful mechanism for funding ICME programs will be public-private partnerships in the form of consortia aimed at solving particularly challenging problems of high priority to the nation. Establishment of international cooperative efforts will be important for accelerating progress and reducing costs. This is a classic cooperation/competition situation in which resources must be leveraged to develop a basic capability while ensuring a competitive and security advantage by being proficient in the application of the tools and by being early adopters. While national security interests may dictate restricting some information on the building blocks of ICME, the committee believes that ensuring as much open access to ICME as possible and making information available to the widest possible audience are also in the best interest of the nation. Such information would include all the elements of the ICME cyberinfrastructure, including collaborative Web sites and repositories of data and models. While export control laws have important purposes, the unnecessary expansion of export control to include ICME could substantially increase the time and cost to develop a widespread ICME capability and limit the ability of multinational corporations to obtain maximum value from it.

Solving the so-called foundational engineering problems in specific high-priority materials systems could provide a focus for the development of ICME and demonstrate its capability. An ICME solution to a key set of engineering problems critical to national security or competitiveness would provide the impetus needed for widespread development and utilization of IMCE. Some possible examples of foundational engineering problems are listed in the report. The committee speculates that based on the case studies identified in Chapter 2, development of an ICME capability for a given material system to solve a particular foundational engineering problem will require an investment of $10 million to $40 million over 3 to 10 years, depending on the completeness and complexity desired. Interestingly, these funding levels are similar to those devoted to genomic challenge problems such as the NIH-funded Rhesus monkey genome project described in Chapter 2. The NNSA effort to develop an ICME capability for nuclear warheads represented a unique situation in that the required experimental efforts were understandably difficult and expensive given the complexities of the materials and systems involved, complexities that led to total costs of more than $150 million. It is important to note that having made an initial investment to develop an ICME capability for a particular foundational engineering problem, this capability can be extended at much less expense. An important feature of these foundational engineering problems for ICME would be the establishment of permanent Web-based cyberinfrastructures to serve as repositories for data and models that would enable future extensions.

The committee concludes that a number of government agencies could play a role. Specific recommendations for these agencies are discussed below and in Chapter 1.

Department of Defense

Since DOD was the sponsor and primary catalyst for one of the central ICME demonstration projects (DARPA's AIM), there is growing and broad awareness of the potential for ICME within DOD. DOD develops and deploys advanced weapons systems of extreme complexity, systems that depend critically on the development of new materials systems and their incorporation into new products.[19,20,21] By providing focused product development objectives as well as long-term sustainable investments, DOD has a unique opportunity to champion the development of a broad ICME capability to enhance and ensure national security. While there will most likely be a slight increase in near-term resource requirements to develop an ICME capability, there is potential for long-term efficiency improvements. In a time of increasingly constrained funding for DOD materials research, ICME provides a means to improve the efficiency of the development of new materials systems, in terms of both cost and time.

The committee has concluded that DOD would benefit from establishing a defense ICME coordination group to champion development of ICME within the defense services and in the industries that supply the services. Because of the many overlaps between the materials needs of the DOD services, a coordinated effort would reap particularly attractive benefits, maximize the value of DOD investments, and minimize the duplication of effort. The tasks for the coordination group would include

- Coordinating and monitoring of DOD ICME efforts that are currently focused on single-agency needs.
- Defining a long-range (15-20 year) strategy and roadmap for the coordinated development of an ICME capability that is specific to the needs of DOD. The strategy would include these:

—Identifying DOD's ICME needs.
—Identifying initial high-priority foundational engineering problems to demonstrate the promise of ICME.
—Establishing and curating a cross-service ICME cyberinfrastructure to include data repositories and libraries of online tools to support the networking and collaboration needed to advance a collaborative ICME culture.

---

[19]Defense Science Board, *Defense Science and Technology*, 2002. Available at http://www.acq.osd.mil/dsb/reports/sandt.pdf. Accessed October 2007.

[20]National Research Council (NRC), *Materials Research to Meet 21st Century Defense Needs*, Washington, D.C.: The National Academies Press (2003).

[21]NRC, *Accelerating Technology Transition: Bridging the Valley of Death for Materials and Processes in Defense Systems*, Washington, D.C.: The National Academies Press (2004).

- Setting policies and establishing procedures to promote public access to data and tools developed from DOD-supported ICME tools, subject to national security concerns.

Department of Energy

ICME has significant potential to provide for the design of new materials for use in energy production and improving the efficiency of its storage and use. The DOE's Office of Energy Efficiency and Renewable Energy (EERE) has championed the ICME discipline, funding the first ICME consortium in the U.S. automotive industry. The consortium was a pilot project for development of an ICME knowledge base and cyberinfrastructure, in this case for magnesium in automotive body applications. The committee concludes that there are likely to be many other areas where EERE can support ICME efforts to develop materials that will impact energy production and energy efficiency. These areas could include efforts in priority areas such as the development of the hydrogen economy, civil infrastructure technologies, solar energy, and vehicle technologies beyond the automobile sector. Moving technology out of the laboratory and into the marketplace is one of EERE's major hurdles, and ICME is a way to achieve that goal.

DOE's Office of Science Basic Energy Sciences (BES) supports a suite of national facilities that could participate in the development of ICME by providing the core fundamental science, computational models, and theory. The committee concludes that there is a particular opportunity for BES to leverage its efforts in computational materials science by linking these theory-based resources to new and yet-to-be-developed rapid characterization techniques for materials, bridging the gaps in theory and computational technique. BES could also accelerate the development of ICME by promoting broad public access to data and tools coming out of federally supported ICME development programs. Here, BES could support coordinated collaborative ICME Web sites which will be used as repositories for the information generated.

DOE's NNSA has played a key role in developing computational materials science tools that predict long-term behavior of selected materials of importance in nuclear weapons systems. NNSA's program on the plutonium life cycle, for example, successfully integrated modeling and experiment across scales, employing a materials modeling approach that ranged from calculations of the electronic structure of materials to large-scale atomistic simulations to determine the long-term properties of microstructural features. While successful in meeting its objectives, the program focused on a single system, and no wider applicability of the ICME process was demonstrated.

The three NNSA laboratories are able to more widely apply an ICME framework to meet their materials development and assessment needs. They are leaders in the development and application of computational materials science, have remarkable computational facilities, and operate a wide-ranging, high-quality experimental program on materials development and characterization. The committee concludes that NNSA laboratories have a unique opportunity to develop such capabilities and to share the ICME framework with the wider materials community.

National Science Foundation

To derive significant benefit from ICME as a new venue for transformational collaborative cross-functional science in materials, physics, and mechanics, as well as to exploit ICME as an efficient mechanism for providing the outputs of fundamental materials research to the engineering community, the committee has concluded that NSF has a significant opportunity to accelerate considerably the development of ICME.[22] The ICME cyberinfrastructure described in Chapter 3 is critical for advancing ICME, and the NSF cyberinfrastructure initiative could be an important source of funding for collaborative, cross-functional international networks in pursuit of NSF's goal of supporting research leading to the development and/or demonstration of innovative cyberinfrastructure services for science and engineering research and education. The protocols and approaches required for

---

[22]NSF has a number of ICME-related programs. For instance, the Center for Computational Materials Design is a collaborative effort between the Pennsylvania State University and the Georgia Institute of Technology and a number of industrial and government sponsors, including the Air Force Research Laboratory, the Army Research Laboratory, Corning, the Ford Motor Company, the General Electric Global Research Center, General Motors, Knolls Atomic Power Laboratory, Lawrence Livermore National Laboratory, Procter & Gamble, Thermo-Calc, and Timken. For more information, see http://www.ccmd.psu.edu/. Accessed February 2008. The nanoHUB was created by the NSF-funded Network for Computational Nanotechnology (NCN), a network of universities with a vision to pioneer the development of nanotechnology from science to manufacturing through innovative theory, exploratory simulation, and novel cyberinfrastructure. The nanoHUB hosts over 790 resources, including online presentations, courses, learning modules, podcasts, animations, teaching materials, and more. Most importantly, the nanoHUB offers simulation tools accessible from a Web browser. For more information, see http://www.nanohub.org/about. Accessed February 2008. NSF's Office of Cyberinfrastructure coordinates and supports the acquisition, development, and provision of state-of-the-art cyberinfrastructure resources. It supports cyberinfrastructure resources, tools, and related services such as supercomputers, high-capacity mass-storage systems, system software suites and programming environments, scalable interactive visualization tools, productivity software libraries and tools, large-scale data repositories and digitized scientific data management systems, networks of various reach and granularity, and an array of software tools and services that hide the complexities and heterogeneity of contemporary cyberinfrastructure while seeking to provide ubiquitous access and enhanced usability. For more information, see http://www.nsf.gov/od/oci/about.jsp. Accessed February 2008.

ICME cyberinfrastructure have yet to be defined, and NSF could play an essential role in helping the ICME community explore this important new area and identify best practices. NSF could also require that all data and models developed during NSF-funded materials research be placed in publicly available ICME cyberinfrastructures. Development of innovative curricula and education for future professionals trained in ICME can also be an important and essential NSF role.

National Institute of Standards and Technology

NIST has a mission to ensure the competitiveness of U.S. industry. ICME holds enormous promise for maintaining and enhancing this competitiveness by providing a highly efficient means to capture materials knowledge and provide it to U.S. manufacturers with a view to optimizing products and manufacturing processes to produce high-quality goods at the lowest possible cost. Material properties in engineered components depend on the manufacturing processes by which they are produced and the manner in which they are used. Thus the data challenges in materials science and engineering are formidable. No single fixed database can be created from which materials engineers can derive the information they need to incorporate materials into a design; instead, there are related databases that can be cross-referenced. This is similar to the situation in bioinformatics, where a diversity of databases must be cross-linked. For example, in the Entrez Genome Project database at the National Center for Biotechnology Information (NCBI), a gene database can be linked to databases on proteins, nucleotides, taxonomy, molecular abundance, three-dimensional structure, and the PubMed database on life sciences journals. Given its unique mission and traditional role as a developer of standardized test techniques and curator of databases, NIST could establish and curate materials informatics databases that can be integrated into ICME tools and collaborative Web sites.

SUMMARY

As discussed in this chapter the committee is convinced that many of the primary barriers to advancing and implementing ICME are cultural and organizational. The time it takes to develop ICME and gain acceptance for it will be directly proportional to the efforts expended in overcoming these cultural barriers. Doing so will remove the major constraint on new industrial products: the absence of a computational materials engineering capability in the product development and optimization cycle.

ICME is the means to integrate materials into the broader computational engineering. Industrial acceptance of ICME is hindered, however, by the slow conversion of science-based materials computational tools to engineering-based tools, by inertia in industry's current product development processes, and by a lack of trained computational materials engineers. To overcome these challenges, industry has to develop an understanding of ICME and its capabilities, and government agencies have a critical key role to play in championing the development of ICME.

The MSE community also has a major role to play. While some aspects of ICME have been successfully implemented, ICME does not exist as a subdiscipline within MSE. For ICME to succeed, it must be embraced as a discipline by the MSE community, the community from which this committee is drawn, and changes in education, research, and information sharing must be brought about. The rate of progress in development of ICME will be proportional as well to the participation of academic researchers in information sharing, model integration, and development of an ICME infrastructure. Materials professional societies have their role to play, too, in establishing the ICME infrastructure, in the continuing education of professionals, and in communicating the progress and successes of ICME through programming and publications.

Appendixes

Appendix A

Committee Membership

Tresa M. Pollock (NAE), *Chair,* is the L.H. and F.E. Van Vlack Professor of Materials Science and Engineering at the University of Michigan, Ann Arbor. She received a B.S. from Purdue University in 1984 and a Ph.D. from MIT in 1989. Dr. Pollock was employed at General Electric Aircraft Engines from 1989 to 1991, where she conducted research and development on high-temperature alloys for aircraft turbine engines. She was a professor in the Department of Materials Science and Engineering at Carnegie Mellon University from 1991 to 1999. Her research interests are in processing and properties of high-temperature structural materials, including nickel-base alloys, intermetallics, coatings, and composites. In 2007 Professor Pollock was president of The Minerals, Metals & Materials Society (TMS), a 10,000-member global technical society for materials professionals, and has served as associate editor of *Metallurgical and Materials Transactions.* She is a fellow of ASM International and has received the ASM International Research Silver Medal Award. Dr. Pollock was elected to the National Academy of Engineering in 2005.

John E. Allison, *Vice Chair,* is a senior technical leader at Ford Research and Advanced Engineering, Ford Motor Company, in Dearborn, Michigan, where he currently leads teams focused on the science and technology required for low-cost, durable components fabricated from cast aluminum and magnesium alloys. Dr. Allison received his Ph.D. in metallurgical engineering and materials science in 1982 from Carnegie Mellon University, his M.S. in metallurgical engineering from Ohio State University in 1977, and his B.S. in engineering mechanics from the U.S. Air Force Academy in 1972. The main focus of Dr. Allison's work is the development of a comprehensive suite of integrated computational materials engineering tools for modeling cast metal components with approaches ranging from casting process simulation to first-principle atomistic calculations. His research expertise is in processing–structure–property relationships, complex failure processes such as fatigue and creep in advanced metals and material selection processes. His past work included the development of titanium, intermetallics, and metal matrix composites for the automotive industry. Dr. Allison has been at Ford Research Laboratories since 1983. His work experience prior to that included service as an officer in the U.S. Air Force at the Wright Aeronautical Laboratories and as a visiting scientist at the Brown-Boveri Corporate Research Center in Baden, Switzerland. Dr. Allison is also an adjunct professor of materials science and engineering at the University of Michigan. He has over 120 publications and 4 patents. In 2002, Dr. Allison was the president of TMS. He is a fellow of ASM and has received numerous awards, including the Arch T. Colwell Award from SAE, Henry Ford Technology Award, Ford Technical Achievement Awards, Ford Innovation Awards, and the Air Force Systems Command Scientific Achievement Award. Dr. Allison has served as a member of NRC's National Materials Advisory Board.

Daniel G. Backman joined the Mechanical Engineering Department of Worcester Polytechnic Institute as a research professor following a 26-year career with GE Aircraft Engines. Dr. Backman received his S.B., S.M., and Sc.D. degrees from the Massachusetts Institute of Technology. He went on to hold an assistant professorship at the University of Illinois at Urbana-Champaign and then joined GE, where he provided materials application engineering support and carried out research on aerospace materials and processes. More recently, he contributed to development of the disk alloy for the NASA high-speed civil transport and led the DARPA-sponsored Accelerated Insertion of Materials (AIM) initiative at GE. Much of Dr. Backman's work has focused on mathematical modeling of material processes and the development and implementation of intelligent processing of materials methods for aircraft engine materials. At the time of his retirement from GE, Dr. Backman was the organizational leader of the Materials Modeling and Simulation section. He has served on a number of national technical committees, a corporate board, and has three patents on aerospace materials.

Mary C. Boyce is the Gail E. Kendall Professor of Mechanical Engineering at the Massachusetts Institute of Technology. Dr. Boyce earned a B.S. degree in engineering science and mechanics from Virginia Tech and an S.M. and a Ph.D. in mechanical engineering from the Massachusetts Institute of Technology. She joined the MIT faculty in 1987. Dr. Boyce teaches mechanics and materials. Her research focuses primarily on the mechanics of elastomers, polymers, and polymeric-based micro- and nanocomposite materials, with emphasis on identifying connections among microstructure, deformation mechanisms, and mechanical properties. She has published over 100 technical journal papers in mechanics and materials. Professor Boyce has received numerous awards and honors recognizing her research and teaching efforts, including the MIT MacVicar Faculty Fellow, the NSF Presidential Young Investigator Award, the ASME Applied Mechanics Young Investigator Award, fellow of the American Academy of Mechanics, fellow of the ASME, and fellow of the American Academy of Arts and Sciences.

Mark Gersh is currently senior manager of the Lockheed Martin Advanced Technology Center's Modeling, Simulation and Information Sciences Department, which is responsible for pursuing destabilizing information technologies critical to the success of the Lockheed Martin Space Systems Company. These technologies are applied in a diverse set of domains, including high-performance imaging and exploitation; mission-specific visualization and interaction; collaborative engineering and optimization; tracking, discrimination, and data fusion; network-embedded autonomy; distributed digital communications; and modeling and simulation integration. Advancements are pursued in the context of remote sensing and space science, telecommunications and navigation, missile defense, space transportation, space exploration, and strategic systems. Previously, Mr. Gersh served as Lockheed Martin's program manager for two key government efforts involved with the advancement of modeling and simulation technologies for rapid design and manufacturing. He led an effort with the Advanced Systems and Technology arm of the National Reconnaissance Office exploring constructs for agile design and development of space systems. He was also the program manager of DARPA's simulation-based design effort, which pioneered the use of virtual prototyping technology in the form of advanced integration frameworks, product modeling techniques, software agent-based services, and multidisciplinary optimization. Prior to joining Lockheed Martin, Mr. Gersh was director of research for the Vanguard Information Technology Strategy Program within Computer Sciences Corporation's subsidiary index. Before that, Mr. Gersh was a program manager for the Information Technology Office at DARPA, where he managed a portfolio of research and advanced technology development that focused on experimental information systems architectures, engineering, and integration. Mr. Gersh received a B.S. in computer engineering from Lehigh University and an M.B.A. from the University of West Florida.

Elizabeth A. Holm is a distinguished member of the technical staff in the Computational Materials Science and Engineering Department at Sandia National Laboratories. She is a computational materials scientist with a long-standing interest in bringing materials modeling to industrial practice. Over her 14 years at Sandia, she has worked on simulations to improve processes that make materials for advanced lighting, on prediction of microcircuit aging and reliability, and on the processing of innovative bearing steels. Her research areas include the theory and modeling of microstructural evolution in complex polycrystals, the physical and mechanical response of microstructures, and the wetting and spreading of liquid metals. Dr. Holm obtained her B.S.E. in materials science and engineering from the University of Michigan, an S.M. in ceramics from the Massachusetts Institute of Technology, and a dual Ph.D. in materials science and engineering and scientific computing from the University of Michigan. She has received several professional honors and awards, is a fellow of ASM International, and serves on the National Materials Advisory Board and the board of directors of TMS. Dr. Holm has authored or coauthored more than 90 publications.

Richard LeSar is professor and chair of the Department of Material Science and Engineering at Iowa State University. He received a B.S. in chemistry from the University of Michigan and an A.M. and a Ph.D. from Harvard University. He spent many years at Los Alamos National Laboratory, serving in a number of research and management positions. Dr. LeSar's work focuses on the development and application of theory, modeling, and simulation of materials structures and properties. He is interested in modeling at many scales, with recent applications of electronic structure calculations (perovskites), atomistic simulations (molecular and metallic systems), and mesoscale simulations (dislocation dynamics). He currently works on employing dislocation simulations to guide the development of new theories of plasticity and the development of coarse-grained descriptions of biomolecules for simulating large-scale molecular processes. Dr. LeSar is a member of the U.S. Air Force Scientific Advisory Board and also serves as a member of the editorial board of the *Annual Reviews of Materials Research*. He is a past editor of *Computational Materials Science*.

Mike Long recently joined the high-performance computing group at Microsoft. Before that, he was a visual supercomputing manager at SGI, where he was responsible for high-performance scientific visualization served to geographically remote users. Prior to SGI, Mr. Long was a senior applications analyst for Linux Networx, where he was responsible for porting and benchmarking manufacturing and process integration and design optimization applications. Prior to joining Linux Networx, Mr. Long was a senior consulting engineering at Engineous Software and a senior applications analyst specializing in high-performance applications for Cray and SGI, where he pioneered many of the optimization efforts, including automotive crashworthiness design optimization and injection molding optimization. Mr. Long received master's and bachelor's degrees in structural analysis from Brigham Young University.

Adam C. Powell IV is principal at Opennovation, an engineering consulting firm in the Boston area. Opennovation provides services in materials process analysis and installation, training, customization, and support of open-source engineering analysis tools. Dr. Powell's technical background is in materials science, with a focus on process technology, including applications in metals, polymers, and thin films. He also has expertise in polymer membranes, electrochemistry, mechanical behavior of materials, fluid mechanics, heat transfer, physical vapor deposition, computer modeling, and high-performance computing and has developed open-source phase field, boundary element, and direct simulation Monte Carlo (DSMC) software. Dr. Powell received S.B. degrees in economics and materials science and engineering, as well as a Ph.D. in materials engineering from the Massachusetts Institute of Technology. Before founding Opennovation, Dr. Powell was a managing engineer at Veryst Engineering LLC. Prior to joining Veryst, Dr. Powell was on the faculty of the Department of Materials Science and Engineering at the Massachusetts Institute of Technology. Before MIT, he was a metallurgist at the National Institute of Standards and Technology.

John (Jack) J. Schirra is the manager of the Engine Systems Integration & Technology Program Office organization in the Materials and Processes Engineering Department at Pratt & Whitney. Prior to that he was manager of the Materials Characterization and Service Investigations groups. He has over 20 years' experience in jet engine materials starting at P&W after graduating from Lehigh University (1984) with a B.S. in materials engineering and metallurgy. While employed at P&W he has also received an M.S. in metallurgy (1987) from Rensselaer Polytechnic Institute and an M.B.A. (2001) from Purdue University. He has over 20 technical publications and presentations and has served on the program committees for both the International Symposium on Superalloys and the Special Emphasis Symposium on Superalloy Inco 718. Throughout most of his career he has worked in structural materials and process development. He has made a significant contribution to the development and utilization of structural materials behavior modeling at P&W starting with integrated empirical superalloy models.

Deborah Demania Whitis is section manager for structural materials development at GE Aviation, responsible for 9 development engineers, 11 materials behavior analysts, and 11 materials testing technicians. Her research group focuses on supporting materials application, development, and repair engineering, as well as life management and design, by developing new structural materials for aircraft engines; conducting specialized materials testing; performing statistical analysis and materials modeling; and publishing materials design curves. She is the department focal point for materials and process modeling tools. Dr. Whitis received a B.S.

in mechanical engineering from the University of Illinois, an M.S. in mechanical engineering from the Massachusetts Institute of Technology, an M.S. in materials science from the University of Cincinnati, and a Ph.D. in materials science and engineering from the University of Virginia. Her academic and industrial experience has involved the development of constitutive models for microstructural evolution and mechanical behavior for high-temperature aerospace alloys. Dr. Whitis served as the modeling task leader for DARPA's AIM program, coordinating the efforts of industry, university, and government laboratory resources to develop multiscale microstructure and property models for nickel-based superalloys. She is a founding member of the Integrated Computational Materials Engineering (ICME) Technical Advisory Group (TAG) for TMS, and she serves as a member of the High-Temperature Alloys Committee of the Structural Materials Division and the Shaping and Forming Committee of the Materials Processing and Manufacturing Division of TMS. She is also a member of the ASME and ASM.

Christopher Woodward is a principal materials research engineer in the Materials and Manufacturing Directorate at the Air Force Research Laboratory. Currently he manages the High-Temperature Metals Development Group, consisting of eight materials scientists and four materials characterization and testing technicians. Areas of research include a wide range of computational materials science and engineering methods ranging from electronic structure and dislocation dynamics to crystal plasticity methods, novel microstructure testing, and physical metallurgy. Dr. Woodward earned a B.S. in physics from the University of Massachusetts and his M.S. and Ph.D. degrees in solid state physics from the University of Illinois. His areas of research include computational materials science, effects of chemistry on alloy thermodynamics and plasticity, electronic structure of complex systems, novel boundary condition methods, dislocation dynamics, and high-performance computing. He has given 25 invited talks at international conferences and produced 55 peer-reviewed publications. Dr. Woodward is the recipient of the Eshbach Fellowship from Northwestern University and the AFOSR Star Team Award (for excellence in basic research) and was an award finalist for the 2005 Charles J. Cleary Scientific Achievement Award at the AFRL.

Appendix B

Acronyms and Abbreviations

AAAS	American Association for the Advancement of Science
AIM	Accelerated Insertion of Materials
ASC	Advanced Simulation and Computing (NNSA program)
ASCI	Advanced Simulation and Computing Initiative
ASCR	Advanced Scientific Computing Research (DOE Office)
BES	Basic Energy Sciences (at the DOE Office of Science)
CAD	computer-aided design
CAE	computer-aided engineering
CALPHAD	CALculation of PHAse Diagram
CFD	computational fluid dynamics
CMS	computational materials science
CT	computed tomography
CTMP	controlled thermomechanical processing of tubes and pipes
DARPA	Defense Advanced Research Projects Agency
DOD	Department of Defense
DOE	Department of Energy
EERE	Office of Energy Efficiency and Renewable Energy (at the Department of Energy)

FD	finite difference
FE	finite element
FEA	finite-element analysis
FOSS	free, open source software
FY	fiscal year
GPU	graphics processing unit
GUI	graphical user interface
HGP	Human Genome Project
HUGO	Human Genome Organization
ICME	integrated computational materials engineering
IHTC	interfacial heat-transfer coefficient
IPD	integrated product development
IPDT	integrated product development team
LANL	Los Alamos National Laboratory
LLNL	Lawrence Livermore National Laboratory
LSF	load-sharing facility
MDO	multidisciplinary optimization
MPI	message-passing interface
MRI	magnetic resonance imaging
MSE	materials science and engineering
NCBI	National Center for Biotechnology Information
NIH	National Institutes of Health
NIST	National Institute of Standards and Technology
NITRD	Networking and Information Technology Research and Development
NNSA	National Nuclear Security Administration
NSF	National Science Foundation
OEM	original equipment manufacturer
ONR	Office of Naval Research
OSG	Open Science Grid
P&W	Pratt & Whitney
PET	positron emission tomography

QMU   Quantification of Margins and Uncertainty

R&D   research and development
ROI   return on investment

SBIR   Small Business Innovation Research
SDM   simulation data manager
SEM   scanning electron microscopy
SSP   Stockpile Stewardship Program (at the NNSA)
STTR   Small Business Technology Transfer

T-ESA   thermal-enhanced spheroidization annealing
TMS   The Minerals, Metals & Materials Society
TOM   tube optimization model

UGT   underground nuclear-explosion testing
UMC   University Materials Council
USAMP   U.S. Automotive Materials Partnership

VAC   virtual aluminum castings
VPP   virtual pilot plant